John Lubbock

The Beauties of Nature and the Wonders of the World we Live in

John Lubbock

The Beauties of Nature and the Wonders of the World we Live in

ISBN/EAN: 9783744715881

Printed in Europe, USA, Canada, Australia, Japan

Cover: Foto ©berggeist007 / pixelio.de

More available books at **www.hansebooks.com**

THE BEAUTIES OF NATURE

GROUP OF BEECHES, BURNHAM.

THE
BEAUTIES OF NATURE

AND THE

WONDERS OF THE WORLD

WE LIVE IN

BY

THE RIGHT HON.
SIR JOHN LUBBOCK, BART., M.P.

F.R.S., D.C.L., LL.D.

New York
MACMILLAN AND CO.
AND LONDON
1892

CONTENTS

CHAPTER I

CHAPTER II

CHAPTER III

CHAPTER IV

CHAPTER V

CHAPTER VI

CHAPTER VII

CHAPTER VIII

CHAPTER IX

CHAPTER X

ILLUSTRATIONS

PLATES

CHAPTER I

INTRODUCTION

B

If any one gave you a few acres, you would say that you had received a benefit; can you deny that the boundless extent of the earth is a benefit? If any one gave you money, you would call that a benefit. God has buried countless masses of gold and silver in the earth. If a house were given you, bright with marble, its roof beautifully painted with colours and gilding, you would call it no small benefit. God has built for you a mansion that fears no fire or ruin . . . covered with a roof which glitters in one fashion by day, and in another by night. . . . Whence comes the breath you draw; the light by which you perform the actions of your life? the blood by which your life is maintained? the meat by which your hunger is appeased? . . . The true God has planted, not a few oxen, but all the herds on their pastures throughout the world, and furnished food to all the flocks; he has ordained the alternation of summer and winter . . . has invented so many arts and varieties of voice, so many notes to make music. . . . We have implanted in us the seed of all ages, of all arts; and God our Master brings forth our intellects from obscurity. — SENECA.

CHAPTER I

INTRODUCTION

THE world we live in is a fairyland of exquisite beauty, our very existence is a miracle in itself, and yet few of us enjoy as we might, and none as yet appreciate fully, the beauties and wonders which surround us. The greatest traveller cannot hope even in a long life to visit more than a very small part of our earth, and even of that which is under our very eyes how little we see!

What we do see depends mainly on what we look for. When we turn our eyes to the sky, it is in most cases merely to see whether it is likely to rain. In the same field the farmer will notice the crop, geologists the fossils, botanists the flowers, artists the colouring, sportsmen the cover for game. Though

3

we may all look at the same things, it does
not at all follow that we should see them.

It is good. as Keble says, " to have our
thoughts lift up to that world where all is
beautiful and glorious," — but it is well to
realise also how much of this world is beautiful.   It has, I know, been maintained. as for
instance by Victor Hugo, that the general
effect of beauty is to sadden. " Comme la
vie de l'homme. même la plus prospère. est
toujours au fond plus triste que gaie. le ciel
sombre nous est harmonieux. Le ciel écla-
tant et joyeux nous est ironique. La Nature
triste nous ressemble et nous console ; la
Nature rayonnante. magnifique. superbe . . .
a quelque chose d'accablant." [1]

This seems to me. I confess. a morbid
view. There are many no doubt on whom
the effect of natural beauty is to intensify
feeling. to deepen melancholy, as well as
to raise the spirits. As Mrs. W. R. Greg
in her memoir of her husband tells us :
" His passionate love for nature. so amply
fed by the beauty of the scenes around him,

[1] *Choses Vues.*

intensified the emotions, as all keen perception of beauty does, but it did not add to their joyousness. We speak of the pleasure which nature and art and music give us; what we really mean is that our whole being is quickened by the uplifting of the veil. Something passes into us which makes our sorrows more sorrowful, our joys more joyful, — our whole life more vivid. So it was with him. The long solitary wanderings over the hills, and the beautiful moonlight nights on the lake served to make the shadows seem darker that were brooding over his home."

But surely to most of us Nature when sombre, or even gloomy, is soothing and consoling; when bright and beautiful, not only raises the spirits, but inspires and elevates our whole being —

> Nature never did betray
> The heart that loved her: 'tis her privilege,
> Through all the years of this our life, to lead
> From joy to joy: for she can so inform
> The mind that is within us, so impress
> With quietness and beauty, and so feed
> With lofty thoughts, that neither evil tongues,
> Rash judgments, nor the sneers of selfish men,

Nor greetings where no kindness is, nor all
The dreary intercourse of daily life,
Shall e'er prevail against us, or disturb
Our cheerful faith, that all which we behold
Is full of blessings.[1]

Kingsley speaks with enthusiasm of the heaths and moors round his home, "where I have so long enjoyed the wonders of nature; never, I can honestly say, alone; because when man was not with me, I had companions in every bee, and flower and pebble; and never idle, because I could not pass a swamp, or a tuft of heather, without finding in it a fairy tale of which I could but decipher here and there a line or two, and yet found them more interesting than all the books, save one, which were ever written upon earth."

Those who love Nature can never be dull. They may have other temptations; but at least they will run no risk of being beguiled, by ennui, idleness, or want of occupation, " to buy the merry madness of an hour with the long penitence of after time." The love of Nature, again, helps us greatly to keep

[1] Wordsworth.

ourselves free from those mean and petty cares which interfere so much with calm and peace of mind. It turns "every ordinary walk into a morning or evening sacrifice," and brightens life until it becomes almost like a fairy tale.

In the romances of the Middle Ages we read of knights who loved, and were loved by, Nature spirits, — of Sir Launfal and the Fairy Tryamour, who furnished him with many good things, including a magic purse, in which

> As oft as thou puttest thy hand therein
> A mark of gold thou shalt iwinne,

as well as protection from the main dangers of life. Such times have passed away, but better ones have come. It is not now merely the few, who are so favoured. All those who love Nature she loves in return, and will richly reward, not perhaps with the good things, as they are commonly called, but with the best things, of this world : not with money and titles, horses and carriages, but with bright and happy thoughts, contentment and peace of mind.

Happy indeed is the naturalist: to him the seasons come round like old friends; to him the birds sing: as he walks along, the flowers stretch out from the hedges, or look up from the ground, and as each year fades away, he looks back on a fresh store of happy memories.

Though we can never "remount the river of our years," he who loves Nature is always young. But what is the love of Nature? Some seem to think they show a love of flowers by gathering them. How often one finds a bunch of withered blossoms on the roadside, plucked only to be thrown away! Is this love of Nature? It is, on the contrary, a wicked waste, for a waste of beauty is almost the worst waste of all.

If we could imagine a day prolonged for a lifetime, or nearly so, and that sunrise and sunset were rare events which happened but a few times to each of us, we should certainly be entranced by the beauty of the morning and evening tints. The golden rays of the morning are a fortune in themselves, but we too often overlook the loveliness of Nature, because it is constantly before us.   For "the senseless folk," says King Alfred,

> is far more struck
> At things it seldom sees.

"Well," says Cicero, "did Aristotle observe, 'If there were men whose habitations had been always underground, in great and commodious houses, adorned with statues and pictures, furnished with everything which they who are reputed happy abound with ; and if, without stirring from thence, they should be informed of a certain divine power and majesty, and, after some time, the earth should open, and they should quit their dark abode to come to us ; where they should immediately behold the earth, the seas, the heavens ; should consider the vast extent of the clouds and force of the winds ; should see the sun, and observe his grandeur and beauty, and also his creative power, inasmuch as day is occasioned by the diffusion of his light through the sky ; and when night has obscured the earth, they should contemplate the heavens bespangled and adorned with stars ; the surprising variety

of the moon, in her increase and wane; the
rising and setting of all the stars, and the
inviolable regularity of their courses; when,
says he, 'they should see these things, they
would undoubtedly conclude that there are
Gods, and that these are their mighty
works.' " [1]

Is my life vulgar, my fate mean,
Which on such golden memories can lean? [2]

At the same time the change which has
taken place in the character of our religion
has in one respect weakened the hold which
Nature has upon our feelings. To the
Greeks — to our own ancestors, — every River
or Mountain or Forest had not only its own
special Deity, but in some sense was itself
instinct with life. They were not only
peopled by Nymphs and Fauns, Elves and
Kelpies, were not only the favourite abodes
of Water, Forest, or Mountain Spirits, but
they had a conscious existence of their own.

In the Middle Ages indeed, these spirits

[1] Cicero, *De Natura Deorum.*
[2] Thoreau.

were regarded as often mischievous, and apt
to take offence; sometimes as essentially
malevolent — even the most beautiful, like
the Venus of Tannhäuser, being often on that
very account all the more dangerous; while
the Mountains and Forests, the Lakes and
Seas, were the abodes of hideous ghosts and
horrible monsters, of Giants and Ogres, Sor-
cerers and Demons. These fears, though
vague, were none the less extreme, and the
judicial records of the Middle Ages furnish
only too conclusive evidence that they were
a terrible reality. The light of Science has
now happily dispelled these fearful nightmares.

Unfortunately, however, as men have mul-
tiplied, their energies have hitherto tended,
not to beautify, but to mar. Forests have
been cut down, and replaced by flat fields in
geometrical squares, or on the continent by
narrow strips. Here and there indeed we
meet with oases, in which beauty has not
been sacrificed to profit, and it is then happily
found that not only is there no loss, but the
earth seems to reward even more richly those
who treat her with love and respect.

Scarcely any part of the world affords so great a variety in so small an area as our own island. Commencing in the south. we have first the blue sea itself, the pebbly beaches, the white chalk cliffs of Kent, the tinted sands of Alum Bay, the Red Sandstone of Devonshire, Granite and Gneiss in Cornwall : inland we have the chalk Downs and clear streams, the well-wooded weald and the rich hop gardens ; farther westwards the undulating gravelly hills, and still farther the granite tors : in the centre of England we have to the east the Norfolk Broads and the Fens ; then the fertile Midlands, the cornfields, rich meadows, and large oxen ; and to the west the Welsh mountains ; farther north the Yorkshire Wolds. the Lancashire hills, the Lakes of Westmoreland ; lastly, the swelling hills, bleak moors, and picturesque castles of Northumberland and Cumberland.

There are of course far larger rivers, but perhaps none lovelier than

> The crystal Thamis wont to glide
> In silver channel, down along the lee,[1]

[1] Spenser.

by lawns and parks, meadows and wooded banks, dotted with country houses and crowned by Windsor Castle itself (see Frontispiece). By many Scotland is considered even more beautiful.

And yet too many of us see nothing in the fields but sacks of wheat, in the meadows but trusses of hay, and in woods but planks for houses, or cover for game. Even from this more prosaic point of view, how much there is to wonder at and admire, in the wonderful chemistry which changes grass and leaves, flowers and seeds, into bread and milk, eggs and cream, butter and honey!

Almost everything, says Hamerton, "that the Peasant does, is lifted above vulgarity by ancient, and often sacred, associations." There is, indeed, hardly any business or occupation with reference to which the same might not be said. The triviality or vulgarity does not depend on what we do, but on the spirit in which it is done. Not only the regular professions, but every useful occupation in life, however humble, is honourable in itself, and may be pursued with dignity and peace.

Working in this spirit we have also the satisfaction of feeling that, as in some mountain track every one who takes the right path, seems to make the way clearer for those who follow; so may we also raise the profession we adopt, and smooth the way for those who come after us. But, even for those who are not Agriculturists, it must be admitted that the country has special charms. One perhaps is the continual change. Every week brings some fresh leaf or flower, bird or insect. Every month again has its own charms and beauty. We sit quietly at home and Nature decks herself for us.

In truth we all love change. Some think they do not care for it, but I doubt if they know themselves.

"Not," said Jefferies, "for many years was I able to see why I went the same round and did not care for change. I do not want change: I want the same old and loved things, the same wild flowers, the same trees and soft ash-green; the turtle-doves, the blackbirds, the coloured yellow-hammer sing, sing, singing so long as there is light to cast a shadow

on the dial, for such is the measure of his
song, and I want them in the same place.
Let me find them morning after morning,
the starry-white petals radiating, striving
upwards up to their ideal. Let me see the
idle shadows resting on the white dust; let
me hear the humble-bees, and stay to look
down on the yellow dandelion disk. Let me
see the very thistles opening their great
crowns — I should miss the thistles; the reed
grasses hiding the moor-hen; the bryony
bine, at first crudely ambitious and lifted by
force of youthful sap straight above the
hedgerow to sink of its weight presently and
progress with crafty tendrils; swifts shot
through the air with outstretched wings like
crescent-headed shaftless arrows darted from
the clouds; the chaffinch with a feather in
her bill; all the living staircase of the spring,
step by step, upwards to the great gallery of
the summer, let me watch the same succession
year by year."

After all then he did enjoy the change
and the succession.

Kingsley again in his charming prose

idyll " My Winter Garden " tries to persuade
himself that he was glad he had never
travelled, " having never yet actually got to
Paris." Monotony, he says, " is pleasant in
itself; morally pleasant, and morally useful.
Marriage is monotonous; but there is much,
I trust, to be said in favour of holy wedlock.
Living in the same house is monotonous;
but three removes, say the wise, are as bad
as a fire. Locomotion is regarded as an evil
by our Litany. The Litany, as usual, is
right. ' Those who travel by land or sea' are
to be objects of our pity and our prayers;
and I do pity them. I delight in that same
monotony. It saves curiosity, anxiety, excitement, disappointment, and a host of bad
passions."

But even as he writes one can see that
he does not convince himself. Possibly, he
admits, " after all, the grapes are sour"; and
when some years after he did travel, how
happy he was! At last, he says, trium-
phantly, " At last we too are crossing the
Atlantic. At last the dream of forty years,
please God, would be fulfilled, and I should

see (and happily not alone), the West Indies and the Spanish Main. From childhood I had studied their Natural History, their Charts, their Romances; and now, at last, I was about to compare books with facts, and judge for myself of the reported wonders of the Earthly Paradise."

No doubt there is much to see everywhere. The Poet and the Naturalist find " tropical forests in every square foot of turf." It may even be better, and especially for the more sensitive natures. to live mostly in quiet scenery. among fields and hedgerows, woods and downs; but it is surely good for every one. from time to time, to refresh and strengthen both mind and body by a spell of Sea air or Mountain beauty.

On the other hand we are told, and told of course with truth, that though mountains may be the cathedrals of Nature. they are generally remote from centres of population; that our great cities are grimy, dark, and ugly; that factories are creeping over several of our counties. blighting them into building ground. replacing trees by chimneys, and destroying almost every vestige of natural beauty.

But if this be true, is it not all the more desirable that our people should have access to pictures and books, which may in some small degree, at any rate, replace what they have thus unfortunately lost? We cannot all travel; and even those who can, are able to see but a small part of the world. Moreover, though no one who has once seen, can ever forget, the Alps, the Swiss lakes, or the Riviera, still the recollection becomes less vivid as years roll on, and it is pleasant, from time to time, to be reminded of their beauties.

There is one other advantage not less important. We sometimes speak as if to visit a country, and to see it, were the same thing. But this is not so. It is not every one who can see Switzerland like a Ruskin or a Tyndall. Their beautiful descriptions of mountain scenery depend less on their mastery of the English language, great as that is, than on their power of seeing what is before them. It has been to me therefore a

matter of much interest to know which
aspects of Nature have given the greatest
pleasure to, or have most impressed, those
who, either from wide experience or from
their love of Nature, may be considered best
able to judge. I will begin with an English
scene from Kingsley. He is describing his
return from a day's trout-fishing : —

"What shall we see," he says, "as we look
across the broad, still, clear river, where the
great dark trout sail to and fro lazily in the
sun? White chalk fields above, quivering
hazy in the heat. A park full of merry hay-
makers ; gay red and blue waggons ; stalwart
horses switching off the flies ; dark avenues
of tall elms ; groups of abele, 'tossing their
whispering silver to the sun' ; and amid them
the house, — a great square red-brick mass,
made light and cheerful though by quoins
and windows of white Sarsden stone, with
high peaked French roofs, broken by louvres
and dormers, haunted by a thousand swallows
and starlings. Old walled gardens, gay with
flowers, shall stretch right and left. Clipt
yew alleys shall wander away into mysterious

glooms, and out of their black arches shall come tripping children, like white fairies, to laugh and talk with the girl who lies dreaming and reading in the hammock there, beneath the black velvet canopy of the great cedar tree, like some fair tropic flower hanging from its boughs; and we will sit down, and eat and drink among the burdock leaves, and then watch the quiet house, and lawn, and flowers, and fair human creatures, and shining water, all sleeping breathless in the glorious light beneath the glorious blue, till we doze off, lulled by the murmur of a thousand insects, and the rich minstrelsy of nightingale and blackcap, thrush and dove.

" Peaceful, graceful, complete English country life and country houses; everywhere finish and polish; Nature perfected by the wealth and art of peaceful centuries! Why should I exchange you, even for the sight of all the Alps ?"

Though Jefferies was unfortunately never able to travel, few men have loved Nature more devotedly, and speaking of his own home he expresses his opinion that : " Of all

sweet things there is none so sweet as fresh
air — one great flower it is, drawn round about,
over, and enclosing us, like Aphrodite's arms;
as if the dome of the sky were a bell-flower
drooping down over us, and the magical
essence of it filling all the room of the earth.
Sweetest of all things is wild-flower air. Full
of their ideal the starry flowers strained upwards on the bank, striving to keep above
the rude grasses that push by them; genius
has ever had such a struggle. The plain road
was made beautiful by the many thoughts it
gave. I came every morning to stay by the
star-lit bank."

Passing to countries across the ocean, Humboldt tells us that: " If I might be allowed to
abandon myself to the recollection of my own
distant travels, I would instance, amongst the
most striking scenes of nature, the calm sub-
limity of a tropical night, when the stars, not
sparkling, as in our northern skies, shed their
soft and planetary light over the gently heav-
ing ocean; or I would recall the deep valleys
of the Cordilleras, where the tall and slender
palms pierce the leafy veil around them, and

waving on high their feathery and arrow-like
branches, form, as it were, ' a forest above a
forest ' ; or I would describe the summit of
the Peak of Teneriffe, when a horizon layer
of clouds, dazzling in whiteness, has separated
the cone of cinders from the plain below, and
suddenly the ascending current pierces the
cloudy veil, so that the eye of the traveller
may range from the brink of the crater, along
the vine-clad slopes of Orotava, to the orange
gardens and banana groves that skirt the
shore.   In scenes like these, it is not the
peaceful charm uniformly spread over the face
of nature that moves the heart, but rather the
peculiar physiognomy and conformation of the
land, the features of the landscape, the ever-
varying outline of the clouds, and their blend-
ing with the horizon of the sea, whether it
lies spread before us like a smooth and shining
mirror, or is dimly seen through the morning
mist.   All that the senses can but imperfectly
comprehend, all that is most awful in such
romantic scenes of nature, may become a
source of enjoyment to man, by opening a wide
field to the creative power of his imagination.

Impressions change with the varying movements of the mind, and we are led by a happy illusion to believe that we receive from the external world that with which we have ourselves invested it."

Humboldt also singles out for especial praise the following description given of Tahiti by Darwin [1] : —

"The land capable of cultivation is scarcely in any part more than a fringe of low alluvial soil, accumulated round the base of mountains, and protected from the waves of the sea by a coral reef, which encircles at a distance the entire line of coast. The reef is broken in several parts so that ships can pass through, and the lake of smooth water within, thus affords a safe harbour, as well as a channel for the native canoes. The low land which comes down to the beach of coral sand is covered by the most beautiful productions of the intertropical regions. In the midst of bananas, orange, cocoa-nut, and breadfruit trees, spots are cleared where yams, sweet potatoes, sugar-cane, and pine-apples are cultivated. Even

[1] Darwin's *Voyage of the Beagle.*

the brushwood is a fruit tree, namely, the guava, which from its abundance is as noxious as a weed. In Brazil I have often admired the contrast of varied beauty in the banana. palm, and orange tree; here we have in addition the breadfruit tree, conspicuous from its large, glossy, and deeply digitated leaf. It is admirable to behold groves of a tree, sending forth its branches with the force of an English Oak, loaded with large and most nutritious fruit. However little on most occasions utility explains the delight received from any fine prospect, in this case it cannot fail to enter as an element in the feeling. The little winding paths. cool from the surrounding shade. led to the scattered houses; and the owners of these everywhere gave us a cheerful and most hospitable reception."

Darwin himself has told us, after going round the world that " in calling up images of the past. I find the plains of Patagonia frequently cross before my eyes; yet these plains are pronounced by all to be most wretched and useless. They are characterised only by negative possessions; without habitations,

without water, without trees, without mountains, they support only a few dwarf plants.
Why then — and the case is not peculiar to
myself — have these arid wastes taken so firm
possession of my mind? Why have not the
still more level, the greener and more fertile
pampas, which are serviceable to mankind,
produced an equal impression? I can scarcely
analyse these feelings, but it must be partly
owing to the free scope given to the imagination. The plains of Patagonia are boundless,
for they are scarcely practicable, and hence
unknown; they bear the stamp of having thus
lasted for ages, and there appears no limit to
their duration through future time. If, as
the ancients supposed, the flat earth was surrounded by an impassable breadth of water,
or by deserts heated to an intolerable excess,
who would not look at these last boundaries
to man's knowledge with deep but ill-de-
fined sensations?"

Hamerton, whose wide experience and
artistic power make his opinion especially
important, says : —

"I know nothing in the visible world that

combines splendour and purity so perfectly as
a great mountain entirely covered with frozen
snow and reflected in the vast mirror of a
lake. As the sun declines, its thousand
shadows lengthen, pure as the cold green
azure in the depth of a glacier's crevasse, and
the illuminated snow takes first the tender
colour of a white rose, and then the flush of a
red one, and the sky turns to a pale malachite
green, till the rare strange vision fades into
ghastly gray, but leaves with you a permanent
recollection of its too transient beauty." [1]

Wallace especially, and very justly, praises
the description of tropical forest scenery given
by Belt in his charming *Naturalist in Nica-
ragua :* —

"On each side of the road great trees
towered up, carrying their crowns out of sight
amongst a canopy of foliage, and with lianas
hanging from nearly every bough, and passing
from tree to tree, entangling the giants in a
great network of coiling cables. Sometimes
a tree appears covered with beautiful flowers
which do not belong to it, but to one of the

[1] Hamerton's *Landscape.*

lianas that twines through its branches and
sends down great rope-like stems to the
ground. Climbing ferns and vanilla cling to
the trunks, and a thousand epiphytes perch
themselves on the branches. Amongst these
are large arums that send down long aerial
roots, tough and strong, and universally used
instead of cordage by the natives. Amongst
the undergrowth several small species of
palms, varying in height from two to fifteen
feet, are common ; and now and then magnificent tree ferns send off their feathery crowns
twenty feet from the ground to delight the
sight by their graceful elegance. Great broad-
leaved heliconias, leathery melastomæ, and
succulent-stemmed, lop-sided leaved and flesh-
coloured begonias are abundant, and typical of
tropical American forests ; but not less so are
the cecropia trees, with their white stems and
large palmated leaves standing up like great
candelabra. Sometimes the ground is carpeted
with large flowers, yellow, pink, or white,
that have fallen from some invisible tree-top
above ; or the air is filled with a delicious
perfume, the source of which one seeks around

in vain, for the flowers that cause it are far overhead out of sight, lost in the great over-shadowing crown of verdure."

" But," he adds, " the uniformity of climate which has led to this rich luxuriance and end-less variety of vegetation is also the cause of a monotony that in time becomes oppressive." To quote the words of Mr. Belt : " Unknown are the autumn tints, the bright browns and yellows of English woods ; much less the crim-sons, purples, and yellows of Canada, where the dying foliage rivals, nay, excels, the ex-piring dolphin in splendour. Unknown the cold sleep of winter ; unknown the lovely awakening of vegetation at the first gentle touch of spring. A ceaseless round of ever-active life weaves the fairest scenery of the tropics into one monotonous whole, of which the component parts exhibit in detail untold variety of beauty."

Siberia is no doubt as a rule somewhat severe and inhospitable, but M. Patrin men-tions with enthusiasm how one day descend-ing from the frozen summits of the Altai, he came suddenly on a view of the plain of the

Obi — the most beautiful spectacle, he says, which he had ever witnessed. Behind him were barren rocks and the snows of winter, in front a great plain, not indeed entirely green, or green only in places, and for the rest covered by three flowers, the purple Siberian Iris, the golden Hemerocallis, and the silvery Narcissus — green, purple, gold, and white, as far as the eye could reach.

Wallace tells us that he himself has derived the keenest enjoyment from his sense of colour : —

" The heavenly blue of the firmament, the glowing tints of sunset, the exquisite purity of the snowy mountains, and the endless shades of green presented by the verdure-clad surface of the earth, are a never - failing source of pleasure to all who enjoy the inestimable gift of sight. Yet these constitute, as it were, but the frame and background of a marvellous and ever-changing picture. In contrast with these broad and soothing tints, we have presented to us in the vegetable and animal worlds an infinite variety of objects adorned with the most beautiful and most

varied hues. Flowers, insects, and birds are
the organisms most generally ornamented in
this way; and their symmetry of form, their
variety of structure, and the lavish abundance with which they clothe and enliven
the earth, cause them to be objects of
universal admiration. The relation of this
wealth of colour to our mental and moral
nature is indisputable. The child and the
savage alike admire the gay tints of flowers,
birds, and insects; while to many of us their
contemplation brings a solace and enjoyment
which is both intellectually and morally
beneficial. It can then hardly excite surprise
that this relation was long thought to afford a
sufficient explanation of the phenomena of colour in nature; and although the fact that —

> Full many a flower is born to blush unseen,
> And waste its sweetness on the desert air,

might seem to throw some doubt on the suffi-
ciency of the explanation. the answer was
easy, — that in the progress of discovery man
would. sooner or later. find out and enjoy
every beauty that the hidden recesses of the
earth have in store for him."

Professor Colvin speaks with special admiration of Greek scenery : —

" In other climates, it is only in particular states of the weather that the remote ever seems so close, and then with an effect which is sharp and hard as well as clear; here the clearness is soft ; nothing cuts or glitters, seen through that magic distance ; the air has not only a new transparency so that you can see farther into it than elsewhere, but a new quality, like some crystal of an unknown water, so that to see into it is greater glory." Speaking of the ranges and promontories of sterile limestone, the same writer observes that their colours are as austere and delicate as the forms. " If here the scar of some old quarry throws a stain, or there the clinging of some thin leafage spreads a bloom, the stain is of precious gold, and the bloom of silver. Between the blue of the sky and the tenfold blue of the sea these bare ranges seem, beneath that daylight, to present a whole system of noble colour flung abroad over perfect forms. And wherever, in the general sterility, you find a little moderate verdure — a little

moist grass, a cluster of cypresses — or whenever your eye lights upon the one wood of the
district, the long olive grove of the Cephissus,
you are struck with a sudden sense of richness,
and feel as if the splendours of the tropics
would be nothing to this."

Most travellers have been fascinated by the
beauty of night in the tropics. Our evenings no doubt are often delicious also, though
the mild climate we enjoy is partly due to the
sky being so often overcast. In parts of the
tropics, however, the air is calm and cloudless throughout nearly the whole of the year.
There is no dew, and the inhabitants sleep on
the house-tops, in full view of the brightness
of the stars and the beauty of the sky, which
is almost indescribable.

" Il faisait," says Bernardin de St. Pierre of
such a scene, " une de ces nuits délicieuses, si
communes entre les tropiques, et dont le plus
abile pinceau ne rendrait pas le beauté. La
lune paraissait au milieu du firmament, en-
tourée d'un rideau de nuages, que ses rayons
dissipaient par degrés. Sa lumière se répan-
dait insensiblement sur les montagnes de l'île

et sur leurs pitons, qui brillaient d'un vert argenté. Les vents retenaient leurs haleines. On entendait dans les bois, au fond des vallées, au haut des rochers, de petits cris, de doux murmures d'oiseaux, qui se caressaient dans leurs nids, réjouis par la clarté de la nuit et la tranquillité de l'air. Tous, jusqu'aux insectes, bruissaient sous l'herbe. Les étoiles étincelaient au ciel, et se réfléchissaient au sein de la mer, qui répétait leurs images tremblantes."

In the Arctic and Antarctic regions the nights are often made quite gorgeous by the Northern Lights or Aurora borealis, and the corresponding appearance in the Southern hemisphere. The Aurora borealis generally begins towards evening, and first appears as a faint glimmer in the north, like the approach of dawn. Gradually a curve of light spreads like an immense arch of yellowish-white hue, which gains rapidly in brilliancy, flashes and vibrates like a flame in the wind. Often two or even three arches appear one over the other. After a while coloured rays dart upwards in divergent pencils, often green below, yellow in the centre, and crimson

D

above, while it is said that sometimes almost
black, or at least very dark violet, rays are
interspersed among the rings of light, and
heighten their effect by contrast. Sometimes
the two ends of the arch seem to rise off the
horizon, and the whole sheet of light throbs
and undulates like a fringed curtain of light;
sometimes the sheaves of rays unite into an
immense cupola; while at others the separate
rays seem alternately lit and extinguished.
Gradually the light flickers and fades away,
and has generally disappeared before the first
glimpse of dawn.

We seldom see the Aurora in the south of
England, but we must not complain; our
winters are mild, and every month has its
own charm and beauty.

In January we have the lengthening days.
" February " the first butterfly.
" March " the opening buds.
" April " the young leaves and
spring flowers.
" May " the song of birds.
" June " the sweet new-mown
hay.

In July   we have the summer flowers.
" August   "   the golden grain.
" September   "   the fruit.
" October   "   the autumn tints.
" November   "   the hoar frost on trees
                                and the pure snow.
" December   "   last not least, the holi-
                                days of Christmas,
                                and the bright fire-
                                side.

It is well to begin the year in January, for we have then before us all the hope of spring.

Oh wind,
If winter comes, can spring be long behind?[1]

Spring seems to revive us all. In the Song of Solomon —

My beloved spake, and said unto me,
Rise up, my love, my fair one, and come away.
For, lo, the winter is past,
The rain is over and gone;
The flowers appear on the earth;
The time of the singing of birds is come,
The voice of the turtle is heard in our land,
The fig tree putteth forth her green figs,
And the vines with the tender grape give a good smell.

[1] Shelley.

" But indeed there are days," says Emerson, " which occur in this climate, at almost any season of the year, wherein the world reaches its perfection, when the air, the heavenly bodies, and the earth make a harmony, as if nature would indulge her offspring. . . . These halcyon days may be looked for with a little more assurance in that pure October weather, which we distinguish by the name of the Indian summer. The day, immeasurably long, sleeps over the broad hills and warm wide fields. To have lived through all its sunny hours, seems longevity enough." Yet does not the very name of Indian summer imply the superiority of the summer itself, — the real, the true summer, " when the young corn is bursting into ear; the awned heads of rye, wheat, and barley, and the nodding panicles of oats, shoot from their green and glaucous stems, in broad, level, and waving expanses of present beauty and future promise. The very waters are strewn with flowers : the buck-bean, the water-violet, the elegant flowering rush, and the queen of the waters, the pure and splendid white lily, invest every stream and lonely mere with grace." [1]

For our greater power of perceiving, and therefore of enjoying Nature, we are greatly indebted to Science. Over and above what is visible to the unaided eye, the two magic tubes, the telescope and microscope, have revealed to us, at least partially, the infinitely great and the infinitely little.

Science, our Fairy Godmother, will, unless we perversely reject her help, and refuse her gifts, so richly endow us, that fewer hours of labour will serve to supply us with the material necessaries of life, leaving us more time to ourselves, more leisure to enjoy all that makes life best worth living.

Even now we all have some leisure, and for it we cannot be too grateful.

"If any one," says Seneca, "gave you a few acres, you would say that you had received a benefit; can you deny that the boundless extent of the earth is a benefit? If a house were given you, bright with marble, its roof beautifully painted with colours and

---

[1] Howitt's *Book of the Seasons.*

gilding, you would call it no small benefit.
God has built for you a mansion that fears
no fire or ruin . . . covered with a roof which
glitters in one fashion by day, and in another
by night. Whence comes the breath which
you draw; the light by which you perform
the actions of your life? the blood by which
your life is maintained? the meat by which
your hunger is appeased? . . . The true God
has planted, not a few oxen, but all the herds
on their pastures throughout the world, and
furnished food to all the flocks; he has ordained the alternation of summer and winter
. . . he has invented so many arts and varieties of voice, so many notes to make music.
. . . We have implanted in us the seeds of
all ages, of all arts; and God our Master
brings forth our intellects from obscurity." [1]

[1] Seneca, *De Beneficiis.*

CHAPTER II

ON ANIMAL LIFE

If thy heart be right, then will every creature be to thee a mirror of life, and a book of holy doctrine.

THOMAS À KEMPIS.

CHAPTER II

ON ANIMAL LIFE

THERE is no species of animal or plant which would not well repay, I will not say merely the study of a day, but even the devotion of a lifetime. Their form and structure, development and habits, geographical distribution, relation to other living beings, and past history, constitute an inexhaustible study.

When we consider how much we owe to the Dog, Man's faithful friend, to the noble Horse, the patient Ox, the Cow, the Sheep, and our other domestic animals, we cannot be too grateful to them; and if we cannot, like some ancient nations, actually worship them, we have perhaps fallen into the other extreme, underrate the sacredness of animal life, and treat them too much like mere machines.

41

Some species, however, are no doubt more interesting than others, especially perhaps those which live together in true communities, and which offer so many traits — some sad, some comical, and all interesting, — which reproduce more or less closely the circumstances of our own life.

The modes of animal life are almost infinitely diversified; some live on land, some in water; of those which are aquatic some dwell in rivers, some in lakes or pools, some on the sea-shore, others in the depths of the ocean. Some burrow in the ground, some find their home in the air. Some live in the Arctic regions, some in the burning deserts; one little beetle (Hydrobius) in the thermal waters of Hammam-Meskoutin, at a temperature of 130°. As to food, some are carnivorous and wage open war; some, more insidious, attack their victims from within; others feed on vegetable food. on leaves or wood, on seeds or fruits; in fact, there is scarcely an animal or vegetable substance which is not the special and favourite food of one or more species. Hence to adapt them to these various require-

ments we find the utmost differences of form
and size and structure. Even the same individual often goes through great changes.

GROWTH AND METAMORPHOSES

The development, indeed, of an animal
from birth to maturity is no mere question
of growth. The metamorphoses of Insects
have long excited the wonder and admiration
of all lovers of nature. They depend to a
great extent on the fact that the little
creatures quit the egg at an early stage of
development, and lead a different life, so
that the external forces acting on them,
are very different from those by which they
are affected when they arrive at maturity. A
remarkable case is that of certain Beetles
which are parasitic on Solitary Bees. The
young lava is very active, with six strong
legs. It conceals itself in some flower, and
when the Bee comes in search of honey, leaps
upon her, but is so minute as not to be perceived. The Bee constructs her cell, stores it

with honey, and lays her egg. At that moment the little larva quits the Bee and jumps on to the egg, which she proceeds gradually to devour. Having finished the egg, she attacks the honey; but under these circumstances the activity which was at first so necessary has become useless; the legs which did such good service are no longer required; and the active slim larva changes into a white fleshy grub, which floats comfortably in the honey with its mouth just below the surface.

Even in the same group we may find great differences. For instance, in the family of Insects to which Bees and Wasps belong, some have grub larvæ, such as the Bee and Ant; some have larvæ like caterpillars. such as the Sawflies; and there is a group of minute forms the larvæ of which live inside the eggs of other insects, and present very remarkable and abnormal forms.

These differences depend mainly on the mode of life and the character of the food.

RUDIMENTARY ORGANS

Such modifications may be called adaptive, but there are others of a different origin that have reference to the changes which the race has passed through in bygone ages. In fact the great majority of animals do go through metamorphoses (many of them as remarkable, though not so familiar as those of insects). but in many cases they are passed through within the egg and thus escape popular observation.   Naturalists who accept the theory of evolution, consider that the development of each individual represents to a certain extent that which the species has itself gone through in the lapse of ages; that every individual contains within itself, so to say, a history of the race.   Thus the rudimentary teeth of Cows. Sheep. Whales, etc. (which never emerge from their sockets), the rudimentary toes of many mammals, the hind legs of Whales and of the Boa-constrictor, which are imbedded in the flesh, the rudimentary collar-bone of the Dog, etc., are in-

dications of descent from ancestors in which
these organs were fully developed. Again,
though used for such different purposes, the
paddle of a Whale, the leg of a Horse and of
a Mole, the wing of a Bird or a Bat, and the
arm of a Man, are all constructed on the same
model, include corresponding bones, and are
similarly arranged. The long neck of the
Giraffe, and the short one of the Whale (if
neck it can be called), contain the same
number of vertebræ.

Even after birth the young of allied species
resemble one another much more than the
mature forms. The stripes on the young
Lion, the spots on the young Blackbird, are
well-known cases; and we find the same law
prevalent among the lower animals, as, for
instance, among Insects and Crustacea. The
Lobster, Crab, Shrimp, and Barnacle are very
unlike when full grown, but in their young
stages go through essentially similar metamorphoses.

No animal is perhaps in this respect more
interesting than the Horse. The skull of a
Horse and that of a Man, though differing so

much, are, says Flower,[1] "composed of exactly the same number of bones, having the same general arrangement and relation to each other. Not only the individual bones, but every ridge and surface for the attachment of muscles, and every hole for the passage of artery or nerve, seen in the one can be traced in the other." It is often said that the Horse presents a remarkable peculiarity in that the canine teeth grow but once. There are, however, in most Horses certain spicules or minute points which are shed before the appearance of the permanent canines, and which are probably the last remnants of the true milk canines.

The foot is reduced to a single toe, representing the third digit, but the second and fourth, though rudimentary, are represented by the splint bones; while the foot also contains traces of several muscles, originally belonging to the toes which have now disappeared, and which " linger as it were behind, with new relations and uses, sometimes in a reduced, and almost, if not quite, function-

[1] *The Horse.*

less condition." Even Man himself presents traces of gill-openings, and indications of other organs which are fully developed in lower animals.

MODIFICATIONS

There is in New Zealand a form of Crow (Hura), in which the female has undergone a very curious modification. It is the only case I know, in which the bill is differently shaped in the two sexes. The bird has taken on the habits of a Woodpecker, and the stout crow-like bill of the cock-bird is admirably adapted to tap trees, and if they sound hollow, to dig down to the burrow of the Insect; but it lacks the horny-pointed tip of the tongue, which in the true Woodpecker is provided with recurved hairs, thus enabling that bird to pierce the grub and draw it out. In the Hura, however, the bill of the hen-bird has become much elongated and slightly curved, and when the cock has dug down to the burrow, the hen inserts her long bill and draws out the grub, which they then divide between them: a very pretty illustration of the wife as helpmate to the husband.

It was indeed until lately the general opinion that animals and plants came into existence just as we now see them. We took pleasure in their beauty; their adaptation to their habits and mode of life in many cases could not be overlooked or misunderstood. Nevertheless the book of Nature was like some missal richly illuminated, but written in an unknown tongue. The graceful forms of the letters, the beauty of the colouring, excited our wonder and admiration; but of the true meaning little was known to us; indeed we scarcely realised that there was any meaning to decipher. Now glimpses of the truth are gradually revealing themselves, we perceive that there is a reason, and in many cases we know what the reason is, for every difference in form, in size, and in colour; for every bone and every feather, almost for every hair.[1]

[1] Lubbock, *Fifty Years of Science.*

COLOUR

The colours of animals, generally, I believe. serve as a protection. In some, however, they probably render them more attractive to their mates, of which the Peacock is one of the most remarkable illustrations.

In richness of colour birds and insects vie even with flowers. "One fine red admiral butterfly," says Jefferies,[1] " whose broad wings, stretched out like fans, looked simply splendid floating round and round the willows which marked the margin of a dry pool. His blue markings were really blue — blue velvet — his red and the white stroke shone as if sunbeams were in his wings. I wish there were more of these butterflies; in summer. dry summer, when the flowers seem gone and the grass is not so dear to us, and the leaves are dull with heat, a little colour is so pleasant. To me colour is a sort of food; every spot of colour is a drop of wine to the spirit."

The varied colours which add so much to

[1] *The Open Air.*

the beauty of animals and plants are not only thus a delight to the eye, but afford us also some of the most interesting problems in Natural History. Some probably are not in themselves of any direct advantage. The brilliant mother-of-pearl of certain shells, which during life is completely hidden, the rich colours of some internal organs of animals, are not perhaps of any direct benefit, but are incidental, like the rich and brilliant hues of many minerals and precious stones.

But although this may be true, I believe that most of these colours are now of some advantage. "The black back and silvery belly of fishes" have been recently referred to by a distinguished naturalist as being obviously of no direct benefit. I should on the contrary have quoted this case as one where the advantage was obvious. The dark back renders the fish less conspicuous to an eye looking down into the water; while the white under-surface makes them less visible from below. The animals of the desert are sand-coloured; those of the Arctic regions are white like snow, especially in winter; and pelagic animals are blue.

Let us take certain special cases. The Lion, like other desert animals, is sand-coloured; the Tiger which lives in the Jungle has vertical stripes, making him difficult to see among the upright grass; Leopards and the tree-cats are spotted, like rays of light seen through leaves.

An interesting case is that of the animals living in the Sargasso or gulf-weed of the Atlantic. These creatures — Fish, Crustacea, and Mollusks alike — are characterised by a peculiar colouring, not continuously olive like the Seaweed itself, but blotched with rounded more or less irregular patches of bright, opake white, so as closely to resemble fronds covered with patches of Flustra or Barnacles.

Take the case of caterpillars, which are especially defenceless, and which as a rule feed on leaves. The smallest and youngest are green, like the leaves on which they live. When they become larger, they are characterised by longitudinal lines, which break up the surface and thus render them less conspicuous. On older and larger ones the lines are diagonal, like the nerves of leaves. Conspicuous caterpillars are generally either nauseous in taste, or protected by hairs.

Fig. 1.— *Chærocampa porcellus.*

I say "generally," because there are some interesting exceptions. The large caterpillars of some of the Elephant Hawkmoths are very conspicuous, and rendered all the more so by the presence of a pair of large eyelike spots. Every one who sees one of these caterpillars is struck by its likeness to a snake, and the so-called "eyes" do much to increase the deception. Moreover, the ring on which they are placed is swollen, and the insect, when in danger, has the habit of retracting its head and front segments, which gives it an additional resemblance to some small reptile. That small birds are, as a matter of fact, afraid of these caterpillars (which, however, I need not say, are in reality altogether harmless) Weis-

mann has proved by actual experiment. He
put one of these caterpillars in a tray, in
which he was accustomed to place seed for
birds. Soon a little flock of sparrows and
other small birds assembled to feed as usual.
One of them lit on the edge of this tray, and
was just going to hop in, when she spied the
caterpillar. Immediately she began bobbing
her head up and down in the odd way which
some small birds have, but was afraid to go
nearer. Another joined her and then another,
until at last there was a little company of ten
or twelve birds all looking on in astonishment,
but not one ventured into the tray; while
one bird, which lit in it unsuspectingly, beat a
hasty retreat in evident alarm as soon as she
perceived the caterpillar. After waiting for
some time, Weismann removed it, when the
birds soon attacked the seeds. Other caterpillars also are probably protected by their
curious resemblance to spotted snakes. One
of the large Indian caterpillars has even acquired the power of hissing.

Among perfect insects many resemble closely
the substances near which they live. Some

moths are mottled so as to mimic the bark of trees, or moss, or the surface of stones. One beautiful tropical butterfly has a dark wing on which are painted a series of green leaf tips, so that it closely resembles the edge of a pinnate leaf projecting out of shade into sunshine.

The argument is strengthened by those cases in which the protection, or other advantage, is due not merely to colour, but partly also to form. Such are the insects which resemble sticks or leaves. Again, there are cases in which insects mimic others, which, for some reason or other, are less liable to danger. So also many harmless animals mimic others which are poisonous or otherwise well protected. Some butterflies, as Mr. Bates has pointed out, mimic others which are nauseous in taste, and therefore not attacked by birds. In these cases it is generally only the females that are mimetic, and in some cases only a part of them, so that there are two, or even three, kinds of females, the one retaining the normal colouring of the group, the other mimicking another species. Some spiders

closely resemble Ants, and several other insects mimic Wasps or Hornets.

Some reptiles and fish have actually the
power of changing the colour of their skin so
as to adapt themselves to their surroundings.

Many cases in which the colouring does not
at first sight appear to be protective, will on
consideration be found to be so. It has, for
instance, been objected that sheep are not
coloured green ; but every mountaineer knows
that sheep could not have had a colour more
adapted to render them inconspicuous, and
that it is almost impossible to distinguish them
from the rocks which so constantly crop up
on hill sides. Even the brilliant blue of the
Kingfisher, which in a museum renders it so
conspicuous, in its native haunts, on the contrary, makes it difficult to distinguish from a
flash of light upon the water ; and the richly-
coloured Woodpecker wears the genuine dress
of a Forester — the green coat and crimson
cap.

It has been found that some brilliantly
coloured and conspicuous animals are either
nauseous or poisonous. In these cases the

brilliant colour is doubtless a protection by rendering them more unmistakable.

COMMUNITIES

Some animals may delight us especially by their beauty, such as birds or butterflies; others may surprise us by their size, as Elephants and Whales, or the still more marvellous monsters of ancient times; may fascinate us by their exquisite forms, such as many microscopic shells; or compel our reluctant attention by their similarity to us in structure; but none offer more points of interest than those which live in communities. I do not allude to the temporary assemblages of Starlings, Swallows, and other birds at certain times of year, nor even to the permanent associations of animals brought together by common wants in suitable localities, but to regular and more or less organised associations. Such colonies as those of Rooks and Beavers have no doubt interesting revelations and surprises in store for us, but they have not been as yet so much studied

as those of some insects. Among these the
Hive Bees, from the beauty and regularity
of their cells, from their utility to man, and
from the debt we owe them for their unconscious agency in the improvement of flowers,
hold a very high place; but they are probably less intelligent, and their relations with
other animals and with one another are less
complex than in the case of Ants, which have
been so well studied by Gould, Huber, Forel,
M·Cook, and other naturalists.

The subject is a wide one, for there are at
least a thousand species of Ants, no two of
which have the same habits. In this country
we have rather more than thirty, most of
which I have kept in confinement. Their life
is comparatively long: I have had working
Ants which were seven years old, and a Queen
Ant lived in one of my nests for fifteen years.
The community consists, in addition to the
young, of males, which do no work, of wingless
workers, and one or more Queen mothers, who
have at first wings, which, however, after one
Marriage flight, they throw off, as they never
leave the nest again, and in it wings would of

course be useless. The workers do not, except occasionally, lay eggs, but carry on all the affairs of the community. Some of them, and especially the younger ones, remain in the nest, excavate chambers and tunnels, and tend the young, which are sorted up according to age, so that my nests often had the appearance of a school, with the children arranged in classes.

In our English Ants the workers in each species are all similar except in size, but among foreign species there are some in which there are two or even more classes of workers, differing greatly not only in size, but also in form. The differences are not the result of age, nor of race, but are adaptations to different functions, the nature of which, however, is not yet well understood. Among the Termites those of one class certainly seem to act as soldiers, and among the true Ants also some have comparatively immense heads and powerful jaws. It is doubtful, however, whether they form a real army. Bates observed that on a foraging expedition the large-headed individuals did not walk in the regular ranks, nor on the return did they carry any of the booty, but marched along at the side, and at tolerably regular intervals, "like subaltern officers in a marching regiment." He is disposed, however, to ascribe to them a much humbler function, namely, to serve merely "as indigestible morsels to the ant thrushes." This, I confess, seems to me improbable.

Solomon was, so far as we yet know, quite correct in describing Ants as having "neither guide, overseer, nor ruler." The so-called Queens are really Mothers. Nevertheless it is true, and it is curious, that the working Ants and Bees always turn their heads towards the Queen. It seems as if the sight of her gave them pleasure. On one occasion, while moving some Ants from one nest into another for exhibition at the Royal Institution, I unfortunately crushed the Queen and killed her. The others, however, did not desert her, or draw her out as they do dead workers, but on the contrary carried her into the new nest, and subsequently into a larger one with which I supplied them, congregating round her for weeks just as if she had been alive. One could hardly help fancying that they were mourning her loss, or hoping anxiously for her recovery.

The Communities of Ants are sometimes very large, numbering even up to 500,000 individuals; and it is a lesson to us, that no one has ever yet seen a quarrel between any two Ants belonging to the same community. On the other hand it must be admitted that they are in hostility, not only with most other insects, including Ants of different species, but even with those of the same species if belonging to different communities. I have over and over again introduced Ants from one of my nests into another nest of the same species, and they were invariably attacked, seized by a leg or an antenna, and dragged out.

It is evident therefore that the Ants of each community all recognise one another, which is very remarkable. But more than this, I several times divided a nest into two halves, and found that even after a separation of a year and nine months they recognised

one another, and were perfectly friendly;
while they at once attacked Ants from a
different nest, although of the same species.

It has been suggested that the Ants of each
nest have some sign or password by which
they recognise one another. To test this I
made some insensible. First I tried chloro-
form, but this was fatal to them; and as
therefore they were practically dead, I did
not consider the test satisfactory. I decided
therefore to intoxicate them. This was
less easy than I had expected. None of
my Ants would voluntarily degrade themselves by getting drunk. However, I got
over the difficulty by putting them into
whisky for a few moments. I took fifty
specimens, twenty-five from one nest and
twenty-five from another, made them dead
drunk, marked each with a spot of paint, and
put them on a table close to where other Ants
from one of the nests were feeding. The
table was surrounded as usual with a moat of
water to prevent them from straying. The
Ants which were feeding soon noticed those
which I had made drunk. They seemed quite

astonished to find their comrades in such a disgraceful condition, and as much at a loss to know what to do with their drunkards as we are. After a while, however, to cut my story short, they carried them all away: the strangers they took to the edge of the moat and dropped into the water, while they bore their friends home into the nest, where by degrees they slept off the effects of the spirit. Thus it is evident that they know their friends even when incapable of giving any sign or password.

This little experiment also shows that they help comrades in distress. If a Wolf or a Rook be ill or injured, we are told that it is driven away or even killed by its comrades. Not so with Ants. For instance, in one of my nests an unfortunate Ant, in emerging from the chrysalis skin, injured her legs so much that she lay on her back quite helpless. For three months, however, she was carefully fed and tended by the other Ants. In another case an Ant in the same manner had injured her antennæ. I watched her also carefully to see what would happen. For some days she did not leave the nest. At last one day she ventured outside, and after a while met a stranger Ant of the same species, but belonging to another nest, by whom she was at once attacked. I tried to separate them, but whether by her enemy, or perhaps by my well-meant but clumsy kindness, she was evidently much hurt and lay helplessly on her side. Several other Ants passed her without taking any notice, but soon one came up, examined her carefully with her antennæ, and carried her off tenderly to the nest. No one, I think, who saw it could have denied to that Ant one attribute of humanity, the quality of kindness.

The existence of such communities as those of Ants or Bees implies, no doubt, some power of communication, but the amount is still a matter of doubt. It is well known that if one Bee or Ant discovers a store of food, others soon find their way to it. This, however, does not prove much. It makes all the difference whether they are brought or sent. If they merely accompany on her return a companion who has brought a store of food,

it does not imply much.   To test this, therefore, I made several experiments.   For instance, one cold day my Ants were almost all
in their nests.   One only was out hunting
and about six feet from home.   I took a dead
bluebottle fly, pinned it on to a piece of cork,
and put it down just in front of her.   She at
once tried to carry off the fly, but to her surprise found it immovable.   She tugged and
tugged, first one way and then another for
about twenty minutes, and then went straight
off to the nest.   During that time not a single
Ant had come out; in fact she was the only
Ant of that nest out at the time.   She went
straight in, but in a few seconds — less than
half a minute, — came out again with no less
than twelve friends, who trooped off with her,
and eventually tore up the dead fly, carrying
it off in triumph.

Now the first Ant took nothing home with
her; she must therefore somehow have made
her friends understand that she had found
some food, and wanted them to come and help
her to secure it.   In all such cases, however.
so far as my experience goes, the Ants brought

their friends, and some of my experiments indicated that they are unable to send them.

Certain species of Ants, again, make slaves of others, as Huber first observed. If a colony of the slave-making Ants is changing the nest, a matter which is left to the discretion of the slaves, the latter carry their mistresses to their new home. Again, if I uncovered one of my nests of the Fuscous Ant (Formica fusca), they all began running about in search of some place of refuge. If now I covered over one small part of the nest, after a while some Ant discovered it. In such a case, however, the brave little insect never remained there, she came out in search of her friends, and the first one she met she took up in her jaws, threw over her shoulder (their way of carrying friends), and took into the covered part; then both came out again, found two more friends and brought them in, the same manœuvre being repeated until the whole community was in a place of safety. This I think says much for their public spirit, but seems to prove that, in F. fusca at least, the powers of communication are but limited.

One kind of slave-making Ant has become so completely dependent on their slaves, that even if provided with food they will die of hunger, unless there is a slave to put it into their mouth. I found, however, that they would thrive very well if supplied with a slave for an hour or so once a week to clean and feed them.

But in many cases the community does not consist of Ants only. They have domestic animals, and indeed it is not going too far to say that they have domesticated more animals than we have. Of these the most important are Aphides. Some species keep Aphides on trees and bushes, others collect root-feeding Aphides into their nests. They serve as cows to the Ants, which feed on the honey-dew secreted by the Aphides. Not only, moreover, do the Ants protect the Aphides themselves, but collect their eggs in autumn, and tend them carefully through the winter, ready for the next spring. Many other insects are also domesticated by Ants, and some of them, from living constantly underground, have completely lost their eyes and become quite blind.

But I must not let myself be carried away by this fascinating subject, which I have treated more at length in another work.[1] I will only say that though their intelligence is no doubt limited, still I do not think that any one who has studied the life-history of Ants can draw any fundamental line of separation between instinct and reason.

When we see a community of Ants working together in perfect harmony, it is impossible not to ask ourselves how far they are mere exquisite automatons: how far they are conscious beings? When we watch an ant-hill tenanted by thousands of industrious inhabitants, excavating chambers, forming tunnels, making roads, guarding their home, gathering food, feeding the young, tending their domestic animals — each one fulfilling its duties industriously, and without confusion, — it is difficult altogether to deny to them the gift of reason; and all our

---

[1] *Ants, Bees, and Wasps.*

recent observations tend to confirm the opinion that their mental powers differ from those of men, not so much in kind as in degree.

CHAPTER III

ON ANIMAL LIFE — *continued*

An organic being is a microcosm — a little universe, formed of a host of self-propagating organisms, inconceivably minute and numerous as the stars of heaven.

DARWIN.

CHAPTER III

WE constantly speak of animals as free. A fish, says Ruskin, " is much freer than a Man; and as to a fly, it is a black incarnation of freedom." It is pleasant to think of anything as free, but in this case the idea is, I fear, to a great extent erroneous. Young animals may frolic and play, but older ones take life very seriously. About the habits of fish and flies, indeed, as yet we know very little. Any one, however, who will watch animals will soon satisfy himself how diligently they work. Even when they seem to be idling over flowers, or wandering aimlessly about, they are in truth diligently seeking for food, or collecting materials for nests. The industry of Bees is proverbial. When collecting honey or pollen

they often visit over twenty flowers in a
minute. keeping constantly to one species,
without yielding a moment's dalliance to any
more sweet or lovely tempter. Ants fully
deserve the commendation of Solomon.
Wasps have not the same reputation for industry; but I have watched them from before
four in the morning till dark at night working like animated machines without a moment's rest or intermission. Sundays and
Bank Holidays are all the same to them.
Again, Birds have their own gardens and
farms from which they do not wander, and
within which they will tolerate no interfer-
ence. Their ideas of the rights of property
are far stricter than those of some statesmen.
As to freedom, they have their daily duties as
much as a mechanic in a mill or a clerk in an
office. They suffer under alarms, moreover.
from which we are happily free. Mr. Galton
believes that the life of wild animals is very
anxious. "From my own recollection," he
says, "I believe that every antelope in South
Africa has to run for its life every one or two
days upon an average, and that he starts or

gallops under the influence of a false alarm many times in a day. Those who have crouched at night by the side of pools in the desert, in order to have a shot at the beasts that frequent it, see strange scenes of animal life; how the creatures gambol at one moment and fight at another; how a herd suddenly halts in strained attention, and then breaks into a maddened rush as one of them becomes conscious of the stealthy movements or rank scent of a beast of prey. Now this hourly life-and-death excitement is a keen delight to most wild creatures, but must be peculiarly distracting to the comfort-loving temperament of others. The latter are alone suited to endure the crass habits and dull routine of domesticated life. Suppose that an animal which has been captured and half-tamed, received ill-usage from his captors, either as punishment or through mere brutality, and that he rushed indignantly into the forest with his ribs aching from blows and stones. If a comfort-loving animal, he will probably be no gainer by the change, more serious alarms and no less ill-usage awaits him: he hears the roar of the wild beasts, and the headlong gallop of the frightened herds, and he finds the buttings and the kicks of other animals harder to endure than the blows from which he fled: he has peculiar disadvantages from being a stranger; the herds of his own species which he seeks for companionship constitute so many cliques, into which he can only find admission by more fighting with their strongest members than he has spirit to undergo. As a set-off against these miseries, the freedom of savage life has no charms for his temperament; so the end of it is, that with a heavy heart he turns back to the habitation he had quitted."

But though animals may not be free, I hope and believe that they are happy. Dr. Hudson, an admirable observer, assures us with confidence that the struggle for existence leaves them much leisure and famous spirits. "In the animal world," he exclaims,[1] "what happiness reigns! What ease, grace, beauty, leisure, and content! Watch these living specks as they glide through their

[1] Address to Microscopical Society, 1890.

forests of algæ, all 'without hurry and care,' as if their 'span-long lives' really could endure for the thousand years that the old catch pines for. Here is no greedy jostling at the banquet that nature has spread for them; no dread of each other; but a leisurely inspection of the field, that shows neither the pressure of hunger nor the dread of an enemy.

" 'To labour and to be content ' (that 'sweet life' of the son of Sirach) — to be equally ready for an enemy or a friend — to trust in themselves alone, to show a brave unconcern for the morrow, all these are the admirable points of a character almost universal among animals, and one that would lighten many a heart were it more common among men. That character is the direct result of the golden law 'If one will not work, neither let him eat'; a law whose stern kindness, unflinchingly applied, has produced whole nations of living creatures, without a pauper in their ranks, flushed with health, alert, resolute, self-reliant, and singularly happy."

It has often been said that Man is the only

animal gifted with the power of enjoying a joke, but if animals do not laugh, at any rate they sometimes play. We are, indeed, apt perhaps to credit them with too much of our own attributes and emotions, but we can hardly be mistaken in supposing that they enjoy certain scents and sounds. It is difficult to separate the games of kittens and lambs from those of children. Our countryman Gould long ago described the "amusements or sportive exercises" which he had observed among Ants. Forel was at first incredulous, but finally confirmed these statements; and, speaking of certain tropical Ants, Bates says "the conclusion that they were engaged in play was irresistible."

SLEEP

We share with other animals the great blessing of Sleep, nature's soft nurse, " the mantle that covers thought, the food that appeases hunger, the drink that quenches thirst, the fire that warms cold, the cold that moderates heat, the coin that purchases all things, the balance and weight that equals the shepherd with the king, and the simple with the wise." Some animals dream as we do; Dogs, for instance, evidently dream of the chase. With the lower animals which cannot shut their eyes it is, however, more difficult to make sure whether they are awake or asleep. I have often noticed insects at night, even when it was warm and light, behave just as if they were asleep, and take no notice of objects which would certainly have startled them in the day. The same thing has also been observed in the case of fish.

But why should we sleep? What a remarkable thing it is that one-third of our life should be passed in unconsciousness. "Half of our days," says Sir T. Browne, " we pass in the shadow of the earth, and the brother of death extracteth a third part of our lives." The obvious suggestion is that we require rest. But this does not fully meet the case. In sleep the mind is still awake, and lives a life of its own: our thoughts wander, uncontrolled, by the will. The mind, therefore, is not necessarily itself at rest; and yet we all know how it is refreshed by sleep.

But though animals sleep, many of them are nocturnal in their habits. Humboldt gives a vivid description of night in a Brazilian forest.

"Everything passed tranquilly till eleven at night, and then a noise so terrible arose in the neighbouring forest that it was almost impossible to close our eyes. Amid the cries of so many wild beasts howling at once the Indians discriminated such only as were (at intervals) heard separately. These were the little soft cries of the sapajous, the moans of the alouate apes, the howlings of the jaguar and couguar, the peccary and the sloth, and the cries of (many) birds. When the jaguars approached the skirt of the forest our dog, which till then had never ceased barking, began to howl and seek for shelter beneath our hammocks. Sometimes, after a long silence, the cry of the tiger came from the tops of the trees; and then it was followed by the sharp and long whistling of the monkeys, which appeared to flee from the danger which threatened them. We heard the same noises repeated during the course of whole months whenever the forest approached the bed of the river.

" When the natives are interrogated on the causes of the tremendous noise made by the beasts of the forest at certain hours of the night, the answer is, they are keeping the feast of the full moon. I believe this agitation is most frequently the effect of some conflict that has arisen in the depths of the forest. The jaguars, for instance, pursue the peccaries and the tapirs, which, having no defence, flee in close troops, and break down the bushes they find in their way. Terrified at this struggle, the timid and distrustful monkeys answer, from the tops of the trees, the cries of the large animals. They awaken the birds that live in society, and by degrees the whole assembly is in commotion. It is not always in a fine moonlight, but more particularly at the time of a storm of violent showers, that this tumult takes place among the wild beasts. ' May heaven grant them a quiet night and repose, and us also!' said the

G

monk who accompanied us to the Rio Negro, when, sinking with fatigue, he assisted in arranging our accommodation for the night."

Life is indeed among animals a struggle for existence, and in addition to the more usual weapons — teeth and claws — we find in some animals special and peculiar means of offence and defence.

If we had not been so familiarised with the fact, the possession of poison might well seem a wonderful gift. That a fluid, harmless in one animal itself, should yet prove so deadly when transferred to others, is certainly very remarkable; and though the venom of the Cobra or the Rattlesnake appeal perhaps more effectively to our imagination, we have conclusive evidence of concentrated poison even in the bite of a midge, which may remain for days perceptible. The sting of a Bee or Wasp, though somewhat similar in its effect, is a totally different organ, being a modified ovipositor. Some species of Ants do not sting in the ordinary sense, but eject their acrid poison to a distance of several inches.

Another very remarkable weapon is the

electric battery of certain Eels, of the Electric
Cat Fish, and the Torpedoes, one of which is
said to be able to discharge an amount of
electricity sufficient to kill a Man.

Some of the Medusæ and other Zoophytes
are armed by millions of minute organs
known as " thread cells." Each consists of a
cell, within which a firm, elastic thread is
tightly coiled. The moment the Medusa
touches its prey the cells burst and the
threads spring out. Entering the flesh as
they do by myriads, they prove very effective
weapons.

The ink of the Sepia has passed into a proverb.
The animal possesses a store of dark fluid,
which, if attacked, it at once ejects, and thus
escapes under cover of the cloud thus created.

The so-called Bombardier Beetles, when attacked, discharge at the enemy, from the
hinder part of their body, an acrid fluid which,
as soon as it comes in contact with air, explodes with a sound resembling a miniature
gun. Westwood mentions, on the authority
of Burchell, that on one occasion, " whilst
resting for the night on the banks of one of

the large South American rivers, he went out
with a lantern to make an astronomical observation, accompanied by one of his black ser-
vant boys; and as they were proceeding,
their attention was directed to numerous
beetles running about upon the shore, which,
when captured, proved to be specimens of a
large species of Brachinus. On being seized
they immediately began to play off their artil-
lery, burning and staining the flesh to such a
degree that only a few specimens could be
captured with the naked hand, and leaving a
mark which remained a considerable time.
Upon observing the whitish vapour with
which the explosions were accompanied, the
negro exclaimed in his broken English, with evident surprise, 'Ah, massa, they make smoke!'"

Many other remarkable illustrations might
be quoted; as for instance the web of the
Spider, the pit of the Ant Lion, the mephitic
odour of the Skunk.

SENSES

We generally attribute to animals five
senses more or less resembling our own. But

even as regards our own senses we really know or understand very little. Take the question of colour. The rainbow is commonly said to consist of seven colours — red, orange, yellow, green, blue, indigo, and violet.

But it is now known that all our colour sensations are mixtures of three simple colours, red, green, and violet. We are, however, absolutely ignorant how we perceive these colours. Thomas Young suggested that we have three different systems of nerve fibres, and Helmholtz regards this as "a not improbable supposition"; but so far as microscopical examination is concerned, there is no evidence whatever for it.

Or take again the sense of Hearing. The vibrations of the air no doubt play upon the drum of the ear, and the waves thus produced are conducted through a complex chain of small bones to the fenestra ovalis and so to the inner ear or labyrinth. But beyond this all is uncertainty. The labyrinth consists mainly of two parts (1) the cochlea, and (2) the semicircular canals, which are three in number, standing at right angles to one

another. It has been supposed that they
enable us to maintain the equilibrium of the
body, but no satisfactory explanation of their
function has yet been given. In the cochlea,
Corti discovered a remarkable organ consisting of some four thousand complex arches,
which increase regularly in length and diminish in height. They are connected at one end
with the fibres of the auditory nerve, and
Helmholtz has suggested that the waves of
sound play on them, like the fingers of a per-
former on the keys of a piano, each separate
arch corresponding to a different sound. We
thus obtain a glimpse, though but a glimpse,
of the manner in which perhaps we hear; but
when we pass on to the senses of smell and
taste, all we know is that the extreme nerve
fibres terminate in certain cells which differ
in form from those of the general surface;
but in what manner the innumerable differences of taste or smell are communicated to
the brain, we are absolutely ignorant.

If then we know so little about ourselves,
no wonder that with reference to other animals our ignorance is extreme.

We are too apt to suppose that the senses of animals must closely resemble, and be confined to ours.

No one can doubt that the sensations of other animals differ in many ways from ours. Their organs are sometimes constructed on different principles, and situated in very unexpected places. There are animals which have eyes on their backs, ears in their legs, and sing through their sides.

We all know that the senses of animals are in many cases much more acute than ours, as for instance the power of scent in the dog, of sight in the eagle. Moreover, our eye is much more sensitive to some colours than to others; least so to crimson, then successively to red, orange, yellow, blue, and green; the sensitiveness for green being as much as 750 times as great as for red. This alone may make objects appear of very different colours to different animals.

Nor is the difference one of degree merely. The rainbow, as we see it, consists of seven colours — red, orange, yellow, green, blue, indigo, and violet. But though the red and violet are the limits of the visible spectrum, they are not the limits of the spectrum itself. there are rays, though invisible to us, beyond the red at the one end, and beyond the violet at the other: the existence of the ultra red can be demonstrated by the thermometer; while the ultra violet are capable of taking a photograph. But though the red and violet are respectively the limits of our vision. I have shown[1] by experiments which have been repeated and confirmed by other naturalists, that some of the lower animals are capable of perceiving the ultra-violet rays, which to us are invisible. It is an interesting question whether these rays may not produce on them the impression of a new colour, or colours, differing from any of those known to us.

So again with hearing, not only may animals in some cases hear better than we do, but sounds which are beyond the reach of our ears, may be audible to theirs. Even among ourselves the power of hearing shrill sounds is greater in some persons than in others. Sound, as we know, is produced by

[1] *Ants, Bees, and Wasps,* and *The Senses of Animals.*

vibration of the air striking on the drum of
the ear, and the fewer are the vibrations in
a second, the deeper is the sound, which
becomes shriller and shriller as the waves of
sound become more rapid. In human ears
the limits of hearing are reached when about
35,000 vibrations strike the drum of the ear
in a second.

Whatever the explanation of the gift of
hearing in ourselves may be, different plans
seem to be adopted in the case of other
animals. In many Crustacea and Insects
there are flattened hairs each connected with
a nerve fibre, and so constituted as to vibrate
in response to particular notes. In others
the ear cavity contains certain minute solid
bodies, known as otoliths, which in the same
way play upon the nerve fibres. Sometimes
these are secreted by the walls of the cavity
itself, but certain Crustacea have acquired the
remarkable habit of selecting after each
moult suitable particles of sand, which they
pick up with their pincers and insert into
their ears.

Many insects, besides the two large

"compound" eyes one on each side of the head, have between them three small ones, known as the "ocelli," arranged in a triangle. The structure of these two sets of eyes is quite different. The ocelli appear to see as our eyes do. The lens throws an inverted image on the back of the eye, so that with these eyes they must see everything reversed, as we ourselves really do, though long practice enables us to correct the impression. On the other hand, the compound eyes consist of a number of facets, in some species as many as 20,000 in each eye, and the prevailing impression among entomologists now is that each facet receives the impression of one pencil of rays, that in fact the image formed in a compound eye is a sort of mosaic. In that case, vision by means of these eyes must be direct; and it is indeed difficult to understand how an insect can obtain a correct impression when it looks at the world with five eyes, three of which see everything reversed, while the other two see things the right way up!

On the other hand, some regard each

facet as an independent eye, in which case
many insects realise the epigram of Plato —

> Thou lookest on the stars, my love,
>   Ah, would that I could be
> Yon starry skies with thousand eyes,
>   That I might look on thee!

Even so, therefore, we only substitute one
difficulty for another.

But this is not all. We have not only no
proof that animals are confined to our five
senses, but there are strong reasons for believing that this is not the case.

In the first place, many animals have
organs which from their position, structure,
and rich supply of nerves, are evidently
organs of sense; and yet which do not
appear to be adapted to any one of our five
senses.

As already mentioned, the limits of hearing
are reached when about 35,000 vibrations
of the air strike on the drums of our ears.
Light, as was first conclusively demonstrated
by our great countryman Young, is the impression produced by vibration of the ether

on the retina of the eye. When 700 millions of millions of vibrations strike the eye in a second. we see violet ; and the colour changes as the number diminishes, 400 millions of millions giving us the impression of red.

Between 35 thousand and 400 millions of millions the interval is immense. and it is obvious that there might be any number of sensations. When we consider how greatly animals differ from us, alike in habits and structure, is it not possible, nay, more, is it not likely that some of these problematical organs are the seats of senses unknown to us, and give rise to sensations of which we have no conception ?

In addition to the capacity for receiving and perceiving, some animals have the faculty of emitting light. In our country the glow-worm is the most familiar case, though some other insects and worms have, at any rate under certain conditions, the same power, and it is possible that many others are really luminous, though with light which is invisible to us. In warmer climates the Fire-fly, Lanthorn-fly, and many other insects, shine with much greater brilliance, and in these cases the glow seems to be a real love-light, like the lamp of Hero.

Many small marine animals, Medusæ. Crustacea, Worms, etc., are also brilliantly luminous at night. Deep-sea animals are endowed also in many cases with special luminous organs, to which I shall refer again.

SENSE OF DIRECTION

It has been supposed that animals possess also what has been called a Sense of Direction. Many interesting cases are on record of animals finding their way home after being taken a considerable distance. To account for this fact it has been suggested that animals possess a sense with which we are not endowed, or of which, at any rate, we possess only a trace. The homing instinct of the pigeon has also been ascribed to the same faculty. My brother Alfred, however, who has paid much attention to pigeons, informs me that they are never taken any great distance at once; but if they are intended to take a long flight, they are trained to do so by stages.

Darwin suggested that it would be interesting to test the case by taking animals in a close box, and then whirling them round rapidly before letting them out. This is in fact done with cats in some parts of France, when the family migrates, and is considered the only way of preventing the cat from returning to the old home. Fabre has tried the same thing with some wild Bees (Chalicodoma). He took some, marked them on the back with a spot of white, and put them into a bag. He then carried them a quarter of a mile, stopping at a point where an old cross stands by the wayside, and whirled the bag rapidly round his head. While he was doing so a good woman came by, who seemed not a little surprised to find the Professor solemnly whirling a black bag round his head in front of the cross; and, he fears, suspected him of Satanic practices. He then carried his Bees a mile and a half in the opposite direction and let them go. Three out of ten found their way home. He tried the same experiment several times, in one case taking them a little over two miles. On an average about a third of the Bees found their way home. "La démonstration," says Fabre, " est suffisante. Ni les mouvements enchevêtrés d'une rotation comme je l'ai décrite ; ni l'obstacle de collines à franchir et de bois à traverser ; ni les embûches d'une voie qui s'avance, rétrograde, et revient par un ample circuit, ne peuvent troubler les Chalicodomes dépaysés et les empêcher de revenir au nid."

I must say, however, that I am not convinced. In the first place, the distances were I think too short ; and in the second, though it is true that some of the Bees found their way home, nearly two-thirds failed to do so. It would be interesting to try the experiment again, taking the Bees say five miles. If they really possess any such sense, that distance would be no bar to their return. I have myself experimented with Ants, taking them about fifty yards from the nest, and I always found that they wandered aimlessly

about, having evidently not the slightest idea
of their way home. They certainly did not
appear to possess any "sense of direction."

NUMBER OF SPECIES

The total number of species may probably
be safely estimated as at least 2,000,000, of
which but a fraction have yet been described
or named. Of extinct species the number
was probably at least as great. In the
geological history of the earth there have
been at least twelve periods, in each of which
by far the greatest number were distinct. The
Ancient Poets described certain gifted mortals
as having been privileged to descend into the
interior of the earth, and exercised their
imagination in recounting the wonders thus
revealed. As in other cases, however, the
realities of Science have proved far more
varied and surprising than the dreams of
fiction. Of these extinct species our knowledge is even more incomplete than that of
the existing species. But even of our contem-

poraries it is not too much to say that, as in the case of plants, there is not one the structure, habits, and life-history of which are yet fully known to us. The male of the Cynips, which produces the common King Charles Oak Apple, has only recently been discovered, those of the root-feeding Aphides, which live in hundreds in every nest of the yellow Meadow Ant (Lasius flavus) are still unknown; the habits and mode of reproduction of the common Eel have only just been discovered; and we may even say generally that many of the most interesting recent discoveries have relation to the commonest and most familiar animals.

IMPORTANCE OF THE SMALLER ANIMALS

Whatever pre-eminence Man may claim for himself, other animals have done far more to affect the face of nature. The principal agents have not been the larger or more intelligent, but rather the smaller, and individually less important, species. Beavers may have dammed up many of the rivers of British Columbia, and turned them into a succession of pools or marshes, but this is a slight matter compared with the action of earthworms and insects[1] in the creation of vegetable soil; of the accumulation of animalcules in filling up harbours and lakes; or of Zoophytes in the construction of coral islands.

Microscopic animals make up in number what they lack in size. Paris is built of Infusoria. The Peninsula of Florida, 78,000 square miles in extent, is entirely composed of coral débris and fragments of shells. Chalk consists mainly of Foraminifera and fragments of shells deposited in a deep sea. The number of shells required to make up a cubic inch is almost incredible. Ehrenberg has estimated that of the Bilin polishing slate which caps the mountain, and has a thickness of forty feet, a cubic inch contains many hundred million shells of Infusoria.

In another respect these microscopic organ-

---

[1] Prof. Drummond (*Tropical Africa*) dwells with great force on the manner in which the soil of Central Africa is worked up by the White Ants.

isms are of vital importance. Many diseases are now known, and others suspected, to be entirely due to Bacteria and other minute forms of life (Microbes), which multiply incredibly, and either destroy their victims, or after a while diminish again in numbers. We live indeed in a cloud of Bacteria. At the observatory of Montsouris at Paris it has been calculated that there are about 80 in each cubic meter of air. Elsewhere, however, they are much more numerous. Pasteur's researches on the Silkworm disease led him to the discovery of Bacterium anthracis, the cause of splenic fever. Microbes are present in persons suffering from cholera, typhus, whooping-cough, measles, hydrophobia, etc., but as to their history and connection with disease we have yet much to learn. It is fortunate, indeed, that they do not all attack us.

In surgical cases, again, the danger of compound fractures and mortification of wounds has been found to be mainly due to the presence of microscopic organisms; and Lister, by his antiseptic treatment which destroys these germs or prevents their access, has greatly diminished the danger of operations, and the sufferings of recovery.

SIZE OF ANIMALS

In the size of animals we find every gradation from these atoms which even in the most powerful microscopes appear as mere points, up to the gigantic reptiles of past ages and the Whales of our present ocean. The horned Ray or Skate is 25 feet in length, by 30 in width. The Cuttle-fishes of our seas. though so hideous as to resemble a bad dream. are too small to be formidable ; but off the Newfoundland coast is a species with arms sometimes 30 feet long, so as to be 60 feet from tip to tip. The body, however, is small in proportion. The Giraffe attains a height of over 20 feet ; the Elephant. though not so tall, is more bulky ; the Crocodile reaches a length of over 20 feet. the Python of 60 feet, the extinct Titanosaurus of the American Jurassic beds. the largest land animal yet known to us. 100 feet in length and 30 in height ; the

Whalebone Whale over 70 feet, Sibbald's
Whale is said to have reached 80–90, which
is perhaps the limit. Captain Scoresby indeed mentions a Rorqual no less than 120
feet in length, but this is probably too great
an estimate.

COMPLEXITY OF ANIMAL STRUCTURE

The complexity of animal structure is even
more marvellous than their mere magnitude.
A Caterpillar contains more than 2000 muscles. In our own body are some 2,000,000
perspiration glands, communicating with the
surface by ducts having a total length of some
10 miles; while that of the arteries, veins,
and capillaries must be very great; the blood
contains millions of millions of corpuscles,
each no doubt a complex structure in itself;
the rods in the retina, which are supposed to
be the ultimate recipient of light, are estimated at 30,000,000; and Meinert has calculated that the gray matter of the brain is
built up of at least 600,000,000 cells. No

verbal description, however, can do justice to the marvellous complexity of animal structure, which the microscope alone, and even that but faintly, can enable us to realise.

LENGTH OF LIFE

How little we yet know of the life-history of Animals is illustrated by the vagueness of our information as to the age to which they live. Professor Lankester[1] tells us that "the paucity and uncertainty of observations on this class of facts is extreme." The Rabbit is said to reach 10 years, the Dog and Sheep 10 –12, the Pig 20, the Horse 30, the Camel 100, the Elephant 200, the Greenland Whale 400 (?): among Birds, the Parrot to attain 100 years, the Raven even more. The Atur Parrot mentioned by Humboldt, talked, but could not be understood, because it spoke in the language of an extinct Indian tribe. It is supposed from their rate of growth that among

[1] Lankester, *Comparative Longevity.* See also Weismann, *Duration of Life.*

Fish the Carp is said to reach 150 years; and a Pike, 19 feet long, and weighing 350 lbs., is said to have been taken in Suabia in 1497 carrying a ring, on which was inscribed, " I am the fish which was first of all put into the lake by the hands of the Governor of the Universe, Frederick the Second, the 5th Oct. 1230." This would imply an age of over 267 years. Many Reptiles are no doubt very long-lived. A Tortoise is said to have reached 500 years. As regards the lower animals, the greatest age on record is that of Sir J. Dalzell's Sea Anemone, which lived for over 50 years. Insects are generally short-lived; the Queen Bee, however, is said by Aristotle, whose statement has not been confirmed by recent writers, to live 7 years. I myself had a Queen Ant which attained the age of 15 years.

The May Fly (Ephemera) is celebrated as living only for a day, and has given its name to all things short-lived. The statement usually made is, indeed, very misleading, for in its larval condition the Ephemera lives for weeks. Many writers have expressed surprise

that in the perfect state its life should be so
short. It is, however, so defenceless, and,
moreover, so much appreciated by birds and
fish, that unless they laid their eggs very
rapidly none would perhaps survive to continue the species.

Many of these estimates are, as will be
seen, very vague and doubtful. so that we
must still admit with Bacon that, " touching
the length and shortness of life in living
creatures, the information which may be had
is but slender, observation is negligent, and
tradition fabulous. In tame creatures their
degenerate life corrupteth them, in wild creatures their exposing to all weathers often in-
tercepteth them."

ON INDIVIDUALITY

When we descend still lower in the animal
scale, the consideration of this question opens
out a very curious and interesting subject
connected with animal individuality. As
regards the animals with which we are most

familiar no such question intrudes. Among quadrupeds and birds, fishes and reptiles, there is no difficulty in deciding whether a given organism is an individual, or a part of an individual. Nor does the difficulty arise in the case of most insects. The Bee or Butterfly lays an egg which develops successively into a larva and pupa, finally producing Bee or Butterfly. In these cases, therefore, the egg, larva, pupa, and perfect Insect, are regarded as stages in the life of a single individual. In certain gnats, however, the larva itself produces young larvæ, each of which develops into a gnat, so that the egg produces not one gnat but many gnats.

The difficulty of determining what constitutes an individual becomes still greater among the Zoophytes. These beautiful creatures in many cases so closely resemble plants, that until our countryman Ellis proved them to be animals, Crabbe was justified in saying —

> Involved in seawrack here we find a race,
> Which Science, doubting, knows not where to place;
> On shell or stone is dropped the embryo seed,
> And quickly vegetates a vital breed.

We cannot wonder that such organisms were long regarded as belonging to the vegetable kingdom. The cups which terminate the branches contain, however, an animal structure, resembling a small Sea Anemone, and possessing arms which capture the food by which the whole colony is nourished. Some of these cups, moreover, differ from the rest, and produce eggs. These then we might be disposed to term ovaries. But in many species they detach themselves from the group and lead an independent existence. Thus we find a complete gradation from structures which, regarded by themselves, we should unquestionably regard as mere organs, to others which are certainly separate and independent beings.

Fig. 2 represents, after Allman, a colony of Bougainvillea fruticosa of the natural size. It is a British species, which is found growing on buoys, floating timber, etc., and. says Allman, " When in health and vigour, offers a spectacle unsurpassed in interest by any other species — every branchlet crowned by its graceful hydranth, and budding with Me-

dusæ in all stages of development (Fig. 3), some
still in the condition of minute buds, in which
no trace of the definite Medusa-form can yet

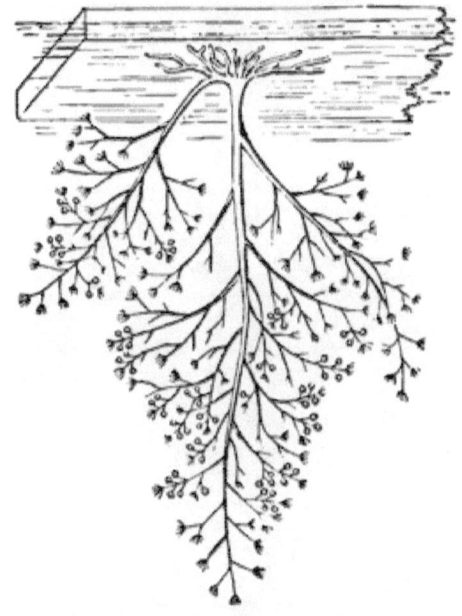

Fig. 2. — Bougainvillea fruticosa; natural size.   (After Allman.)

be detected ; others, in which the outlines of
the Medusa can be distinctly traced within
the transparent ectotheque (external layer) ;
others, again, just casting off this thin outer
pellicle, and others completely freed from it,
struggling with convulsive efforts to break
loose from the colony, and finally launched

forth in the full enjoyment of their freedom
into the surrounding water. I know of no

Fig. 3.—Bougainvillea fruticosa; magnified to show development.

form in which so many of the characteristic
features of a typical hydroid are more finely
expressed than in this beautiful species."

Fig. 4 represents the Medusa or free form of this beautiful species.

If we pass to another great group of Zoophytes, that of the Jelly-fishes, we have a very similar case. For our first knowledge of the life-history of these Zoophytes we are indebted to the Norwegian naturalist Sars. Take, for instance, the common Jelly-fish (Medusa aurita) (Fig. 5) of our shores.

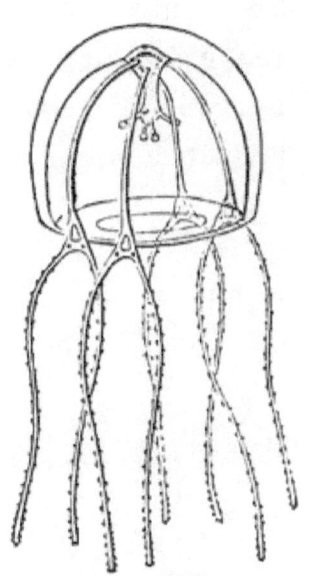

Fig. 4. — Bougainvillea fruticosa, Medusa-form.

The egg is a pear-shaped body (*1*), covered with fine hairs, by the aid of which it swims about, the broader end in front. After a while it attaches itself, not as might have been expected by the posterior but by the anterior extremity (*2*). The cilia then disappear, a mouth is formed at the free end, tentacles, first four (*3*), then eight, and at length as many as thirty (*4*), are formed, and the little creature resembles in essentials the freshwater polyp (Hydra) of our ponds.

At the same time transverse wrinkles (*4*)
are formed round the body, first near the
free extremity and then gradually descend-
ing. They become deeper and deeper, and
develop lobes or divisions one under the other,

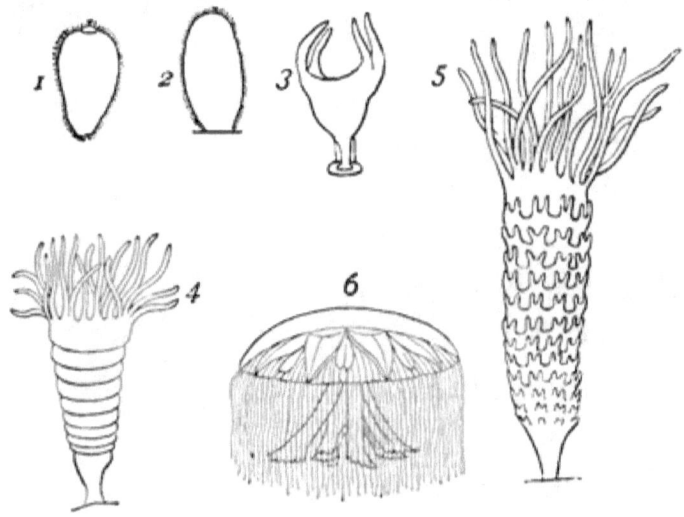

Fig. 5. — Medusa aurita, and progressive stages of development.

as at *5*. After a while the top ring (and
subsequently the others one by one) detaches
itself, swims away, and gradually develops
into a Medusa (*6*). Thus, then, the life-history is very similar to that of the Hydroids.
only that while in the Hydroids the fixed
condition is the more permanent, and the free

swimming more transitory, in the Medusæ, on the contrary, the fixed condition is apparently only a phase in the production of the free swimming animal. In both the one and the other, however, the egg gives rise not to one but to many mature animals. Steenstrup has given to these curious phenomena, many other cases of which occur among the lower animals, and to which he first called attention, the name of alternations of generations.

In the life-history of Infusoria (so called because they swarm in most animal or vegetable infusions) similar difficulties encounter us. The little creatures, many of which are round or oval in form, from time to time become constricted in the middle; the constriction becomes deeper and deeper, and at length the two halves twist themselves apart and swim away. In this case, therefore, there was one, and there are now two exactly similar; but are these two individuals? They are not parent and offspring — that is clear, for they are of the same age; nor are they twins, for there is no parent. As already mentioned, we regard the Caterpillar, Chrysalis, and Butterfly as stages in the life-history of a single individual. But among Zoophytes, and even among some insects, one larva often produces several mature forms. In some species these mature forms remain attached to the larval stock, and we might be disposed to regard the whole as one complex organism. But in others they detach themselves and lead an independent existence.

These considerations then introduce much difficulty into our conception of the idea of an Individual.

ANIMAL IMMORTALITY

But, further than this, we are confronted by by another problem. If we regard a mass of coral as an individual because it arises by continuous growth from a single egg, then it follows that some corals must be thousands of years old.

Some of the lower animals may be cut into pieces, and each piece will develop into an

entire organism. In fact the realisation of
the idea of an individual gradually becomes
more and more difficult, and the continuity of
existence, even among the highest animals,
gradually forces itself upon us. I believe
that as we become more rational, as we realise more fully the conditions of existence,
this consideration is likely to have important
moral results.

It is generally considered that death is the
common lot of all living beings. But is this
necessarily so? Infusoria and other unicellu-
lar animals multiply by division. That is to
say, if we watch one for a certain time, we
shall observe, as already mentioned, that a
constriction takes place, which grows gradually deeper and deeper, until at last the two
halves become quite detached, and each
swims away independently. The process is
repeated over and over again, and in this
manner the species is propagated. Here obviously there is no birth and no death. Such
creatures may be killed, but they have no
natural term of life. They are, in fact, theo-

retically immortal. Those which lived millions of years ago may have gone on dividing and subdividing, and in this sense multitudes of the lower animals are millions of years old.

CHAPTER IV

ON PLANT LIFE

Flower in the crannied wall,
I pluck you out of the crannies,
I hold you here, root and all, in my hand,
Little flower — but *if* I could understand
What you are, root and all, and all in all,
I should know what God and man is.

TENNYSON.

CHAPTER IV

ON PLANT LIFE

WE are told that in old days the Fairies used to give presents of Flowers and Leaves to those whom they wished to reward, or whom they loved best; and though these gifts were, it appears, often received with disappointment, still it will probably be admitted that flowers have contributed more to the happiness of our lives than either gold or silver or precious stones; and that our happiest days have been spent out-of-doors in the woods and fields, when we have

> . . . found in every woodland way
> The sunlight tint of Fairy Gold.[1]

To many minds Flowers acquired an additional interest when it was shown that

[1] Thomson.

there was a reason for their colour, size, and form — in fact, for every detail of their organisation. If we did but know all that the smallest flower could tell us, we should have solved some of the greatest mysteries of Nature. But we cannot hope to succeed — even if we had the genius of Plato or Aristotle — without careful, patient, and reverent study. From such an inquiry we may hope much ; already we have glimpses, enough to convince us that the whole history will open out to us conceptions of the Universe wider and grander than any which the Imagination alone would ever have suggested.

Attempts to explain the forms, colours, and other characteristics of animals and plants are by no means new. Our Teutonic forefathers had a pretty story which explained certain points about several common plants. Balder, the God of Mirth and Merriment, was, characteristically enough, regarded as deficient in the possession of immortality. The other divinities, fearing to lose him, petitioned Thor to make him immortal, and the prayer was granted on condition that every animal and plant would swear not to injure him. To secure this object, Nanna, Balder's wife, descended upon the earth. Loki, the God of Envy, followed her, disguised as a crow (which at that time were white), and settled on a little blue flower, hoping to cover it up, so that Nanna might overlook it. The flower, however, cried out "forget-me-not, forget-me-not," and has ever since been known under that name. Loki then flew up into an oak and sat on a mistletoe. Here he was more successful. Nanna carried off the oath of the oak, but overlooked the mistletoe. She thought, however, and the divinities thought, that she had successfully accomplished her mission, and that Balder had received the gift of immortality.

One day, supposing Balder proof, they amused themselves by shooting at him, posting him against a Holly. Loki tipped an arrow with a piece of Mistletoe, against which Balder was not proof, and gave it to Balder's brother. This, unfortunately, pierced him to the heart, and he fell dead. Some drops of his blood spurted on to the Holly, which

accounts for the redness of the berries; the
Mistletoe was so grieved that she has ever
since borne fruit like tears; and the crow,
whose form Loki had taken, and which till
then had been white, was turned black.

This pretty myth accounts for several things,
but is open to fatal objections.

Recent attempts to explain the facts of
Nature are not less fascinating, and, I think,
more successful.

Why then this marvellous variety? this
inexhaustible treasury of beautiful forms?
Does it result from some innate tendency in
each species? Is it intentionally designed to
delight the eye of man? Or has the form
and size and texture some reference to the
structure and organisation, the habits and
requirements of the whole plant?

I shall never forget hearing Darwin's paper
on the structure of the Cowslip and Primrose,
after which even Sir Joseph Hooker compared
himself to Peter Bell, to whom

> A primrose by a river's brim
> A yellow primrose was to him,
> And it was nothing more.

We all, I think, shared the same feeling, and found that the explanation of the flower then given, and to which I shall refer again, invested it with fresh interest and even with new beauty.

A regular flower, such, for instance, as a Geranium or a Pink, consists of four or more whorls of leaves, more or less modified : the lowest whorl is the Calyx, and the separate leaves of which it is composed, which however are sometimes united into a tube, are called sepals; (2) a second whorl, the corolla, consisting of coloured leaves called petals, which, however, like those of the Calyx, are often united into a tube ; (3) of one or more stamens, consisting of a stalk or filament, and a head or anther, in which the pollen is produced ; and (4) a pistil, which is situated in the centre of the flower, and at the base of which is the Ovary, containing one or more seeds.

Almost all large flowers are brightly coloured, many produce honey, and many are sweet-scented.

What, then, is the use and purpose of this complex organisation ?

It is, I think, well established that the main object of the colour, scent, and honey of flowers is to attract insects, which are of use to the plant in carrying the pollen from flower to flower.

In many species the pollen is, and no doubt it originally was in all, carried by the air. In these cases the chance against any given grain of pollen reaching the pistil of another flower of the same species is of course very great, and the quantity of pollen required is therefore immense.

In species where the pollen is wind-borne as in most of our trees — firs, oaks, beech, ash, elm, etc., and many herbaceous plants, the flowers are as a rule small and inconspicuous, greenish, and without either scent or honey. Moreover, they generally flower early, so that the pollen may not be intercepted by the leaves, but may have a better chance of reaching another flower. And they produce an immense quantity of pollen, as otherwise there would be little chance that any would reach the female flower. Every one must have noticed the clouds of pollen produced by

the Scotch Fir. When, on the contrary, the
pollen is carried by insects, the quantity necessary is greatly reduced. Still it has been
calculated that a Peony flower produces between 3,000,000 and 4,000,000 pollen grains;
in the Dandelion, which is more specialised,
the number is reduced to about 250,000;
while in such a flower as the Dead-nettle it is
still smaller.

The honey attracts the insects; while the
scent and colour help them to find the flowers,
the scent being especially useful at night,
which is perhaps the reason why evening
flowers are so sweet.

It is to insects, then, that flowers owe
their beauty, scent, and sweetness. Just as
gardeners, by continual selection, have added
so much to the beauty of our gardens, so to
the unconscious action of insects is due the
beauty, scent, and sweetness of the flowers of
our woods and fields.

Let us now apply these views to a few
common flowers. Take, for instance, the
White Dead-nettle.

The corolla of this beautiful and familiar

flower (Fig. 6) consists of a narrow tube, somewhat expanded at the upper end (Fig. 7), where the lower lobe forms a platform, on each side of which is a small projecting tooth (Fig. 8, *m*). The upper portion of the corolla is an arched hood (*co*), under which lie four anthers (*a a*), in pairs, while between them, and projecting somewhat downwards, is the pointed pistil (*st*); the tube at the lower part contains honey, and above the honey is a row of hairs running round the tube.

Fig. 6. — White Dead-nettle.

Now, why has the flower this peculiar form? What regulates the length of the tube? What is the use of the arch? What lesson do the little teeth teach us? What advantage is the honey to the flower? Of what use is the fringe of hairs? Why does the stigma project beyond the

anthers? Why is the corolla white, while
the rest of the plant is green?

The honey of course serves to attract the
Humble Bees by which the flower is fertilised,
and to which it is especially adapted; the

Fig. 7.

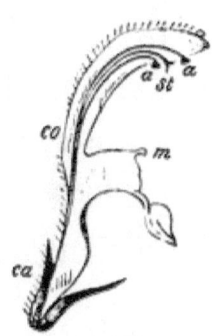

Fig. 8.

white colour makes the flower more conspicuous; the lower lip forms the stage on which
the Bees may alight; the length of the tube
is adapted to that of their proboscis; its
narrowness and the fringe of fine hairs exclude
small insects which might rob the flower of
its honey without performing any service in
return; the arched upper lip protects the
stamens and pistil, and prevents rain-drops
from choking up the tube and washing away
the honey; the little teeth are, I believe, of

no use to the flower in its present condition, they are the last relics of lobes once much larger, and still remaining so in some allied species, but which in the Dead-nettle, being no longer of any use, are gradually disappearing; the height of the arch has reference to the size of the Bee, being just so much above the alighting stage that the Bee, while sucking the honey, rubs its back against the hood and thus comes in contact first with the stigma and then with the anthers, the pollen-grains from which adhere to the hairs on the Bee's back, and are thus carried off to the next flower which the Bee visits, when some of them are then licked off by the viscid tip of the stigma.[1]

In the Salvias. the common blue Salvia of our gardens, for instance,— a plant allied to the Dead-nettle, — the flower (Fig. 9) is constructed on the same plan. but the arch is much larger, so that the back of the Bee does not nearly reach it. The stamens, however, have undergone a remarkable modification. Two of them have become small and function-

---

[1] Lubbock. *Flowers and Insects.*

less. In the other two the anthers or cells producing the pollen, which in most flowers form
together a round knob or
head at the top of the
stamen, are separated by
a long arm, which plays
on the top of the stamen
as on a hinge. Of these
two arms one hangs down
into the tube, closing the
passage, while the other

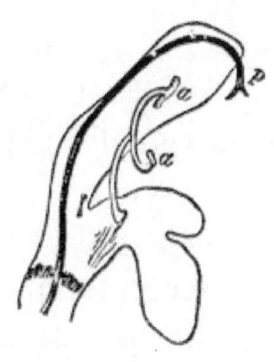

Fig. 9.

lies under the arched upper lip. When the
Bee pushes its proboscis down the tube (Fig. 11)

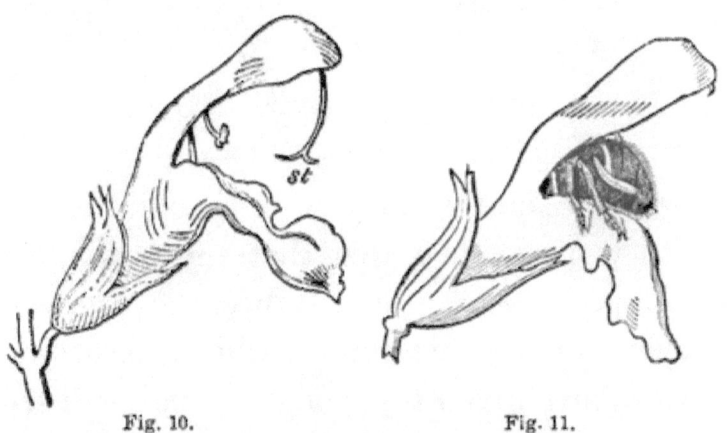

Fig. 10.   Fig. 11.

it presses the lower arm to one side, and the
upper arm consequently descends, tapping the

Bee on the back, and dusting it with pollen. When the flower is a little older the pistil (Fig. 9, p) has elongated so that the stigma (Fig. 10, st) touches the back of the Bee and carries off some of the pollen. This sounds a little complicated, but is clear enough if we take a twig or stalk of grass and push it down the tube, when one arm of each of the two larger stamens will at once make its appearance. It is one of the most beautiful pieces of plant mechanism which I know, and was first described by Sprengel, a poor German schoolmaster.

SNAPDRAGON

At first sight it may seem an objection to the view here advocated that the flowers in some species — as, for instance, the common Snapdragon (Antirrhinum), which, according to the above given tests, ought to be fertilised by insects — are entirely closed. A little consideration, however, will suggest the reply. The Snapdragon is especially adapted for fertilisation by Humble Bees. The stamens and pistil are so arranged that smaller species would not effect the object. It is therefore an advantage that they should be excluded, and in fact they are not strong enough to move the spring. The Antirrhinum is, so to speak, a closed box, of which the Humble Bees alone possess the key.

FURZE, BROOM, AND LABURNUM

Other flowers such as the Furze, Broom, Laburnum, etc., are also opened by Bees. The petals lock more or less into one another, and the flower remains at first closed. When, however, the insect alighting on it presses down the keel, the flower bursts open, and dusts it with pollen.

SWEET PEA

In the above cases the flower once opened does not close again. In others, such as the Sweet Pea and the Bird's-foot Lotus, Nature

K

has been more careful. When the Bee alights it clasps the "wings" of the flower with its legs, thus pressing them down; they are, however, locked into the "keel," or lower petal, which accordingly is also forced down, thus exposing the pollen which rubs against, and part of which sticks to, the breast of the Bee. When she leaves the flower the keel and wings rise again, thus protecting the rest of the pollen and keeping it ready until another visitor comes. It is easy to carry out the same process with the fingers.

PRIMULA

In the Primrose and Cowslip, again, we find quite a different plan. It had long been known that if a number of Cowslips or Primroses are examined, about half would be found to have the stigma at the top of the tube and the stamens half way down, while in the other half the stamens are at the top and the stigma half way down. These two forms are about equally numerous, but never occur on the

same stock. They have been long known to
children and gardeners, who call them thrum-
eyed and pin-eyed. Mr. Darwin was the
first to explain the significance of this curious
difference. It cost him several years of
patient labour, but when once pointed out it
is sufficiently obvious. An insect thrusting its

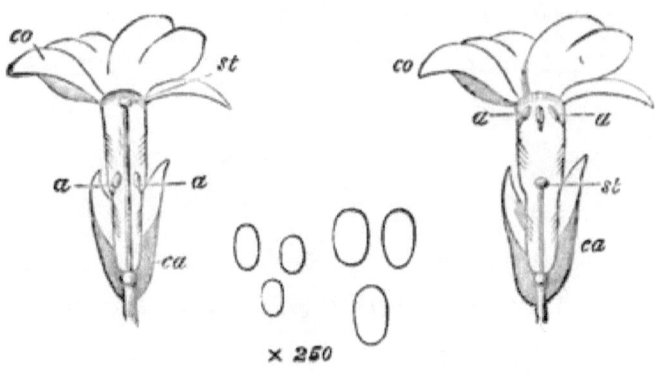

Fig. 12.   Fig. 13.

Flower and Pollen of Primrose

proboscis down a primrose of the long-styled
form (Fig. 12) would dust its proboscis at a
part (*a*) which, when it visited a short-styled
flower (Fig. 13), would come just opposite
the head of the pistil (*st*), and could not fail
to deposit some of the pollen on the stigma.
Conversely, an insect visiting a short-styled
plant would dust its proboscis at a part farther

from the tip; which, when the insect subsequently visited a long-styled flower, would again come just opposite to the head of the pistil. Hence we see that by this beautiful arrangement insects must carry the pollen of the long-styled form to the short-styled. and *vice versâ*.

The economy of pollen is not the only advantage which plants derive from these visits of Insects. A second and scarcely less important is that they tend to secure " cross fertilisation "; that is to say, that the seed shall be fertilised by pollen from another plant. The fact that " cross fertilisation " is of advantage to the plant doubtless also explains the curious arrangement that in many plants the stamen and pistil do not mature at the same time — the former having shed their pollen before the pistil is mature ; or. which happens less often, the pistil having withered before the pollen is ripe. In most Geraniums. Pinks, etc., for instance, and many allied species, the stamens ripen first, and are followed after an interval by the pistil.

THE NOTTINGHAM CATCHFLY

The Nottingham Catchfly (Silene nutans) is a very interesting case. The flower is adapted to be fertilised by Moths. Accordingly it opens towards evening, and as is generally the case with such flowers, is pale in colour, and sweet-scented. There are two sets of stamens, five in each set. The first evening that the flower opens one set of stamens ripen and expose their pollen. Towards morning these wither away, the flower shrivels up, ceases to emit scent, and looks as if it were faded. So it remains all next day. Towards evening it reopens, the second set of stamens have their turn, and the flower again becomes fragrant. By morning, however, the second set of stamens have shrivelled, and the flower is again asleep. Finally on the third evening it re-opens for the last time, the long spiral stigmas expand, and can hardly fail to be fertilised with the pollen brought by Moths from other flowers.

THE HEATH

In the hanging flowers of Heaths the stamens form a ring, and each one bears two
horns. When the Bee inserts its proboscis
into the flower to reach the honey, it is sure
to press against one of these horns, the ring
is dislocated, and the pollen falls on to the
head of the insect. In fact, any number of
other interesting cases might be mentioned.

BEES AND FLIES

Bees are intelligent insects, and would soon
cease to visit flowers which did not supply
them with food. Flies, however, are more
stupid, and are often deceived. Thus in our
lovely little Parnassia, five of the ten stamens
have ceased to produce pollen, but are prolonged into fingers, each terminating in a
shining yellow knob, which looks exactly like
a drop of honey, and by which Flies are con-

tinually deceived. Paris quadrifolia also
takes them in with a deceptive promise of the
same kind. Some foreign plants have livid
yellow and reddish flowers, with a most offen-
sive smell, and are constantly visited by Flies,
which apparently take them for pieces of
decaying meat.

The flower of the common Lords
and Ladies (Arum) of our hedges
is a very interesting case. The
narrow neck bears a number of
hairs pointing downwards. The
stamens are situated above the
stigma, which comes to maturity
first. Small Flies enter the flower
apparently for shelter, but the hairs
prevent them from returning, and
they are kept captive until the
anthers have shed their pollen.
Then, when the Flies have been
well dusted, the hairs shrivel up, leaving a
clear road, and the prisoners are permitted
to escape. The tubular flowers of Aristolochia
offer a very similar case.

Fig. 14.—Arum.

PAST HISTORY OF FLOWERS

If the views here advocated are correct, it follows that the original flowers were small and green, as wind-fertilised flowers are even now. But such flowers are inconspicuous. Those which are coloured, say yellow or white, are of course much more visible and more likely to be visited by insects. I have elsewhere given my reasons for thinking that under these circumstances some flowers became yellow, that some of them became white, others subsequently red, and some finally blue. It will be observed that red and blue flowers are as a rule highly specialised, such as Aconites and Larkspurs as compared with Buttercups; blue Gentians as compared with yellow, etc. I have found by experiment that Bees are especially partial to blue and pink.

Tubular flowers almost always, if not always, contain honey, and are specially suited to Butterflies and Moths, Bees and Flies. Those which are fertilised by Moths generally come out in the evening, are often very sweetly scented, and are generally white or pale yellow, these colours being most visible in the twilight.

Aristotle long ago noticed the curious fact that in each journey Bees confine themselves to some particular flower. This is an economy of labour to the Bee, because she has not to vary her course of proceeding. It is also an advantage to the plants, because the pollen is carried from each flower to another of the same species, and is therefore less likely to be wasted.

FRUITS AND SEEDS

After the flower comes the seed, often contained in a fruit, and which itself encloses the future plant. Fruits and seeds are adapted for dispersion, beautifully and in various ways : some by the wind, being either provided with a wing, as in the fruits of many trees — Sycamores, Ash. Elms, etc. ; or with a hairy crown or covering, as with Thistles, Dandelions, Willows, Cotton plant, etc.

Some seeds are carried by animals ; either
as food — such as most edible fruits and seeds,
acorns, nuts, apples, strawberries, raspberries,
blackberries, plums, grasses, etc. — or invol-
untarily, the seeds having hooked hairs or
processes, such as burrs, cleavers, etc.

Some seeds are scattered by the plants
themselves, as, for instance, those of many
Geraniums, Violets, Balsams, Shamrocks, etc.
Our little Herb Robert throws its seeds some
25 feet.

Some seeds force themselves into the
ground, as those of certain grasses, Cranes'-
bills (Erodiums), etc.

Some are buried by the parent plants,
as those of certain clovers, vetches, violets,
etc.

Some attach themselves to the soil, as
those of the Flax ; or to trees, as in the case
of the Mistletoe.

LEAVES

Again, as regards the leaves there can, I
think, be no doubt that similar considerations

of utility are applicable. Their forms are almost infinitely varied. To quote Ruskin's vivid words, they "take all kinds of strange shapes, as if to invite us to examine them. Star-shaped, heart-shaped, spear-shaped, arrow-shaped, fretted, fringed, cleft, furrowed, serrated, sinuated, in whorls, in tufts, in spires, in wreaths, endlessly expressive, deceptive, fantastic, never the same from foot-stalk to blossom, they seem perpetually to tempt our watchfulness and take delight in outstepping our wonder."

But besides these differences of mere form, there are many others : of structure, texture, and surface; some are scented or have a strong taste, or acrid juice, some are smooth, others hairy; and the hairs again are of various kinds.

I have elsewhere[1] endeavoured to explain some of the causes which have determined these endless varieties. In the Beech, for instance (Fig. 15), the leaf has an area of about 3 square inches. The distance between the buds is about $1\frac{1}{4}$ inch, and the leaves lie in

[1] *Flowers, Fruits, and Leaves.*

the general plane of the branch, which bends slightly at each internode.   The basal half of

the leaf fits the swell of the twig, while the upper half follows the edge of the leaf above ; and the form of the inner edge being thus determined, decides that of the outer one also.

The weight, and consequently the size of the leaf, is limited by the strength of the twig ; and, again, in a climate such as

Fig. 15. — Beech.

ours it is important to plants to have their leaves so arranged as to secure the maximum of light.   Hence in leaves which lie parallel to the plane of the boughs, as in the Beech, the width depends partly on the distance between the buds; if the leaves were broader, they would overlap, if they were narrower, space would be wasted. Consequently the width being determined by the distance between the buds, and the size depending on the weight

which the twig can safely support, the length
also is determined. This argument is well
illustrated by comparing the leaves of the
Beech with those of the Spanish Chestnut.
The arrangement is similar, and the distance
between the buds being about the same, so is
the width of the leaves. But the terminal
branches of the Spanish Chestnut being much
stronger, the leaves can safely be heavier;
hence the width being fixed, they grow in
length and assume the well-known and
peculiar sword-blade shape.

In the Sycamores, Maples (Fig. 16), and
Horse-Chestnuts the arrangement is altogether
different. The shoots are stiff and upright
with leaves placed at right angles to the
branches instead of being parallel to them.
The leaves are in pairs and decussate with
one another; while the lower ones have long
petioles which bring them almost to the level
of the upper pairs, the whole thus forming a
beautiful dome.

For leaves arranged as in the Beech the
gentle swell at the base is admirably suited;
but in a crown of leaves such as those of the

Sycamore, space would be wasted, and it is better that they should expand at once, so soon as their stalks have carried them free from the upper and inner leaves.

In the Black Poplar the arrangement of the leaves is again quite different. The leaf stalk is flattened, so that the leaves hang

Fig. 16. — Acer platanoides.

vertically. In connection with this it will be observed that while in most leaves the upper and under surfaces are quite unlike, in the Black Poplar on the contrary they are very similar. The stomata or breathing holes, moreover, which in the leaves of most trees are confined to the under surface, are in this species nearly equally numerous on both.

The "Compass" Plant of the American prairies, a plant not unlike a small sunflower, is another species with upright leaves, which growing in the wide open prairies tend to point north and south, thus exposing both surfaces equally to the light and heat. Such a position also affects the internal structure of the leaf, the two sides becoming similar in structure, while in other cases the upper and under surfaces are very different.

In the Yew the leaves are inserted close to one another, and are linear; while in the Box they are further apart and broader. In other cases the width of the leaves is determined by what botanists call the "Phyllotaxy." Some plants have the leaves opposite, each pair being at right angles with the pairs above and below.

In others they are alternate, and arranged round the stem in a spiral. In one very common arrangement the sixth leaf stands directly over the first, the intermediate ones forming a spiral which has passed twice round the stem. This, therefore, is known as the $\frac{2}{5}$ arrangement. Common cases are $\frac{1}{2}$, $\frac{1}{3}$, $\frac{2}{5}$, $\frac{3}{8}$, and $\frac{5}{13}$. In the first the leaves are generally broad, in the $\frac{2}{5}$ arrangement they are elliptic, in the $\frac{5}{13}$ and more complicated arrangements nearly linear. The Willows afford a very interesting series. Salix herbacea has the $\frac{1}{3}$ arrangement and rounded leaves, Salix caprea elliptic leaves and $\frac{2}{5}$, Salix pentandra lancet-shaped leaves and $\frac{3}{8}$, and S. incana linear leaves and a $\frac{5}{13}$ arrangement. The result is that whether the series consists of 2, 3, 5, 8, or 13 leaves, in every case, if we look perpendicularly at a twig the leaves occupy the whole circle.

In herbaceous plants upright leaves as a rule are narrow, which is obviously an advantage, while prostrate ones are broad.

AQUATIC PLANTS

Many aquatic plants have two kinds of leaves : some more or less rounded, which float on the surface : and others cut up into narrow segments, which remain below. The latter thus present a greater extent of surface.

AQUATIC VEGETATION, BRAZIL.

To face page 145.

In air such leaves would be unable even to support their own weight, much less to resist the force of the wind. In still air, however, for the same reason, finely-divided leaves may be an advantage, while in exposed positions compact and entire leaves are more suitable. Hence herbaceous plants tend to have divided, bushes and trees entire, leaves. There are many cases when even in the same family low and herb-like species have finely-cut leaves, while in shrubby or ligneous ones they more or less resemble those of the Laurel or Beech.

These considerations affect trees more than herbs, because trees stand more alone, while herbaceous plants are more affected by surrounding plants. Upright leaves tend to be narrow, as in the case of grasses; horizontal leaves, on the contrary, wider. Large leaves are more or less broken up into leaflets, as in the Ash, Mountain-Ash, Horse-Chestnut, etc.

The forms of leaves depend also much on the manner in which they are packed into the buds.

The leaves of our English trees, as I have already said, are so arranged as to secure the maximum of light; in very hot countries the reverse is the case. Hence, in Australia, for instance, the leaves are arranged not horizontally, but vertically, so as to present, not their surfaces, but their edges, to the sun. One English plant, a species of lettuce, has the same habit. This consideration has led also to other changes. In many species the leaves are arranged directly under, so as to shelter, one another. The Australian species of Acacia have lost their true leaves, and the parts which in them we generally call leaves are in reality vertically-flattened leaf stalks.

In other cases the stem itself is green, and to some extent replaces the leaves. In our common Broom we see an approach to this, and the same feature is more marked in Cactus. Or the leaves become fleshy, thus offering, in proportion to their volume, a smaller surface for evaporation. Of this the Stonecrops, Mesembryanthemum, etc., are familiar instances. Other modes of checking transpiration and thus adapting plants to dry situations are by the development of hairs, by the formation of chalky excretions, by the sap becoming saline or viscid, by the leaf becoming more or less rolled up, or protected by a covering of varnish.

Our English trees are for the most part deciduous. Leaves would be comparatively useless in winter when growth is stopped by the cold; moreover, they would hold the snow, and thus cause the boughs to be broken down. Hence perhaps the glossiness of Evergreen leaves, as, for instance, of the Holly, from which the snow slips off. In warmer climates trees tend to retain their leaves, and some species which are deciduous in the north become evergreen, or nearly so, in the south of Europe. Evergreen leaves are as a rule tougher and thicker than those which drop off in autumn; they require more protection from the weather. But some evergreen leaves are much longer lived than others; those of the Evergreen Oak do not survive a second year, those of the Scotch Pine live for three, of the Spruce Fir, Yew, etc., for eight or ten, of the

Pinsapo even eighteen. As a general rule the Conifers with short leaves keep them on for several years, those with long ones for fewer, the length of the leaf being somewhat in the inverse ratio to the length of its life ; but this is not an invariable criterion, as other circumstances also have to be taken into consideration.

Leaves with strong scent, aromatic taste, or acrid juice, are characteristic of dry regions, where they run especial danger of being eaten. and where they are thus more or less effectively protected.

ON HAIRS

The hairs of plants are useful in various ways. In some cases (1) they keep off superfluous moisture; in others (2) they prevent too rapid evaporation ; in some (3) they serve as a protection against too glaring light ; in some (4) they protect the plant from browsing quadrupeds ; in others (5) from being eaten by insects ; or, (6) serve as a quickset hedge to prevent access to the flowers.

In illustration of the first case I may refer
to many alpine plants, the well-known Edel-
weiss, for instance, where the woolly covering
of hairs prevents the " stomata," or minute
pores leading into the interior of the leaf,
from being clogged up by rain, dew, or fog,
and thus enable them to fulfil their functions
as soon as the sun comes out.

As regards the second case many desert and
steppe-plants are covered with felty hairs,
which serve to prevent too rapid evaporation
and consequent loss of moisture.

The woolly hairy leaves of the Mulleins
(Verbascum) doubtless tend to protect them
from being eaten, as also do the spines of
Thistles, and those of Hollies, which, be it
remarked, gradually disappear on the upper
leaves which browsing quadrupeds cannot
reach.

I have already alluded to the various ways
in which flowers are adapted to fertilisation
by insects. But Ants and other small creeping insects cannot effectually secure this object.
Hence it is important that they should be excluded, and not allowed to carry off the honey,

for which they would perform no service in return. In many cases, therefore, the opening of the flower is either contracted to a narrow passage, or is itself protected by a fringe of hairs. In others the peduncle, or the stalk of the plant, is protected by a hedge, or chevaux de frise, of hairs.

In this connection I might allude to the many plants which are more or less viscid. This also is in most cases a provision to preclude creeping insects from access to the flowers.

There are various other kinds of hairs to which I might refer — glandular hairs, secretive hairs, absorbing hairs, etc. It is marvellous how beautifully the form and structure of leaves is adapted to the habits and requirements of the plants, but I must not enlarge further on this interesting subject.

The time indeed will no doubt come when we shall be able to explain every difference of form and structure, almost infinite as these differences are.

INFLUENCE OF SOIL

The character of the vegetation is of course greatly influenced by that of the soil. In this respect granitic and calcareous regions offer perhaps the best marked contrast.

There are in Switzerland two kinds of Rhododendrons, very similar in their flowers, but contrasted in their leaves : Rhododendron hirsutum having them hairy at the edges as the name indicates ; while in R. ferrugineum they are rolled, but not hairy, at the edges, and become ferrugineous on the lower side. This species occurs in the granitic regions, where R. hirsutum does not grow.

The Yarrows (Achillea) afford us a similar case. Achillea atrata and A. moschata will live either on calcareous or granitic soil, but in a district where both occur, A. atrata grows so much the more vigorously of the two if the soil is calcareous that it soon exterminates A. moschata; while in granite districts, on the contrary, A. moschata is victorious and A. atrata disappears.

Every keen sportsman will admit that a varied "bag" has a special charm, and the botanist in a summer's walk may see at least a hundred plants in flower, all with either the interest of novelty, or the charm of an old friend.

ON SEEDLINGS

In many cases the Seedlings afford us an interesting insight into the former condition of the plant. Thus the leaves of the Furze are reduced to thorns; but those of the Seedling are herbaceous and trifoliate like those of the Herb Genet and other allied species, subsequent ones gradually passing into spines. This is evidence that the ancestors of the Furze bore leaves.

Plants may be said to have their habits as well as animals.

SLEEP OF PLANTS

Many flowers close their petals during rain; the advantage of which is that it prevents the honey and pollen from being spoilt

or washed away. Everybody, however, has
observed that even in fine weather certain
flowers close at particular hours. This habit
of going to sleep is surely very curious. Why
should flowers do so? In animals we can
better understand it; they are tired and
require rest. But why should flowers sleep?
Why should some flowers do so, and not
others? Moreover, different flowers keep
different hours. The Daisy opens at sunrise
and closes at sunset, whence its name " day's-
eye.' The Dandelion (Leontodon) is said to
open about seven and to close about five;
Arenaria rubra to be open from nine to three;
the White Water Lily (Nymphæa), from about
seven to four; the common Mouse-ear Hawkweed (Hieracium) from eight to three; the
Scarlet Pimpernel (Anagallis) to waken at
seven and close soon after two; Tragopogon
pratensis to open at four in the morning,
and close just before twelve, whence its
English name, " John go to bed at noon."
Farmers' boys in some parts are said to regulate their dinner time by it. Other flowers,
on the contrary, open in the evening.

Now it is obvious that flowers which are fertilised by night-flying insects would derive no advantage from being open by day; and on the other hand, that those which are fertilised by bees would gain nothing by being open at night. Nay it would be a distinct disadvantage, because it would render them liable to be robbed of their honey and pollen, by insects which are not capable of fertilising them. I have ventured to suggest then that the closing of flowers may have reference to the habits of insects, and it may be observed also in support of this, that wind-fertilised flowers do not sleep; and that many of those flowers which attract insects by smell, open and emit their scent at particular hours; thus Hesperis matronalis and Lychnis vespertina smell in the evening, and Orchis bifolia is particularly sweet at night.

But it is not the flowers only which "sleep" at night; in many species the leaves also change their position, and Darwin has given strong reasons for considering that the object is to check transpiration and thus tend to a protection against cold.

BEHAVIOUR OF LEAVES IN RAIN

The behaviour of plants with reference to rain affords many points of much interest. The Germander Speedwell (Veronica) has two strong rows of hairs, the Chickweed (Stellaria) one, running down the stem and thus conducting the rain to the roots. Plants with a main tap-root, like the Radish or the Beet, have leaves sloping inwards so as to conduct the rain towards the axis of the plant, and consequently to the roots ; while, on the contrary, where the roots are spreading the leaves slope outwards.

In other cases the leaves hold the rain or dew drops. Every one who has been in the Alps must have noticed how the leaves of the Lady's Mantle (Alchemilla) form little cups containing each a sparkling drop of icy water. Kerner has suggested that owing to these cold drops, the cattle and sheep avoid the leaves.

MIMICRY

In many cases plants mimic others which are better protected than themselves. Thus Matricaria Chamomilla mimics the true Chamomile, which from its bitterness is not eaten by quadrupeds. Ajuga Chamæpitys mimics Euphorbia Cyparissias, with which it often grows, and which is protected by its acrid juice. The most familiar case, however, is that of the Stinging and the Dead Nettles. They very generally grow together, and though belonging to quite different families are so similar that they are constantly mistaken for one another. Some Orchids have a curious resemblance to insects, after which they have accordingly been named the Bee Orchis, Fly Orchis, Butterfly Orchis, etc., but it has not yet been satisfactorily shown what advantage the resemblance is to the plant.

ANTS AND PLANTS

The transference of pollen from plant to

plant is by no means the only service which insects render.

Ants, for instance, are in many cases very useful to plants. They destroy immense numbers of caterpillars and other insects. Forel observing a large Ants' nest counted more than 28 insects brought in as food per minute. In some cases Ants attach themselves to particular trees, constituting a sort of bodyguard. A species of Acacia, described by Belt, bears hollow thorns, while each leaflet produces honey in a crater-formed gland at the base, as well as a small, sweet, pear-shaped body at the tip. In consequence it is inhabited by myriads of a small ant, which nests in the hollow thorns, and thus finds meat, drink, and lodging all provided for it. These ants are continually roaming over the plant, and constitute a most efficient bodyguard, not only driving off the leaf-eating ants, but, in Belt's opinion, rendering the leaves less liable to be eaten by herbivorous mammalia. Delpino mentions that on one occasion he was gathering a flower of Clerodendrum, when he was himself suddenly attacked by a whole army of small ants.

INSECTIVOROUS PLANTS

In the cases above mentioned the relation
between flowers and insects is one of mutual
advantage. But this is by no means an invariable rule. Many insects, as we all know,
live on plants, but it came upon botanists as a
surprise when our countryman Ellis first discovered that some plants catch and devour insects. This he observed in a North American
plant Dionæa, the leaves of which are formed
something like a rat-trap, with a hinge in the
middle, and a formidable row of spines round
the edge. On the surface are a few very sensitive hairs, and the moment any small insect
alights on the leaf and touches one of these
hairs the two halves of the leaf close up
quickly and catch it. The surface then throws
out a glutinous secretion, by means of which
the leaf sucks up the nourishment contained
in the insect.

Our common Sun-dews (Drosera) are also
insectivorous, the prey being in their case

captured by glutinous hairs. Again, the Bladderwort (Utricularia), a plant with pretty yellow flowers, growing in pools and slow streams, is so called because it bears a great number of bladders or utricles, each of which is a real miniature eel-trap, having an orifice guarded by a flap opening inwards which allows small water animals to enter, but prevents them from coming out again. The Butterwort (Pinguicula) is another of these carnivorous plants.

MOVEMENTS OF PLANTS

While considering Plant life we must by no means confine our attention to the higher orders, but must remember also those lower groups which converge towards the lower forms of animals, so that in the present state of our knowledge the two cannot always be distinguished with certainty. Many of them differ indeed greatly from the ordinary conception of a plant. Even the comparatively highly organised Seaweeds multiply by means of bodies called spores, which an untrained observer would certainly suppose to be animals. They are covered by vibratile hairs or " cilia," by means of which they swim about freely in the water, and even possess a red spot which, as being especially sensitive to light, may be regarded as an elementary eye, and with the aid of which they select some suitable spot, to which they ultimately attach themselves.

It was long considered as almost a characteristic of plants that they possessed no power of movement. This is now known to be an error. In fact, as Darwin has shown, every growing part of a plant is in continual and even constant rotation. The stems of climbing plants make great sweeps, and in other cases, when the motion is not so apparent, it nevertheless really exists. I have already mentioned that many plants change the position of their leaves or flowers, or, as it is called, sleep at night.

The common Dandelion raises its head when the florets open, opens and shuts morning and evening, then lies down again while the seeds are ripening, and raises itself a second time when they are ready to be carried away by the wind.

Valisneria spiralis is a very interesting case. It is a native of European rivers, and the female flower has a long spiral stalk which enables it to float on the surface of the water. The male flowers have no stalks, and grow low down on the plant. They soon, however, detach themselves altogether, rise to the surface, and thus are enabled to fertilise the female flowers among which they float. The spiral stalk of the female flower then contracts and draws it down to the bottom of the water so that the seeds may ripen in safety. Many plants throw or bury their seeds.

The sensitive plants close their leaves when touched, and the leaflets of Desmodium gyrans are continually revolving. I have already mentioned that the spores of seaweeds swim freely in the water by means of cilia. Some microscopic plants do so throughout a great part of their lives.

A still lower group, the Myxomycetes, which resemble small, more or less branched, masses of jelly, and live in damp soil, among

decaying leaves, under bark and in similar
moist situations, are still more remarkably
animal like. They are never fixed, but in
almost continual movement, due to differences
of moisture, warmth, light, or chemical action.
If, for instance, a moist body is brought into
contact with one of their projections, or
"pseudopods," the protoplasm seems to roll
itself in that direction, and so the whole
organism gradually changes its place. So
again, while a solution of salt, carbonate of
potash, or saltpetre causes them to withdraw
from the danger, an infusion of sugar, or tan,
produces a flow of protoplasm towards the
source of nourishment. In fact, in the same
way it rolls over and round its food, absorbing
what is nutritious as it passes along. In cold
weather they descend into the soil, and one
of them (Œthalium), which lives in tan pits,
descends in winter to a depth of several
feet. When about to fructify it changes its
habits, seeks the light instead of avoiding it,
climbs upwards, and produces its fruit above
ground.

IMPERFECTION OF OUR KNOWLEDGE

The total number of living species of plants may be roughly estimated at 500,000, and there is not one, of which we can say that the structure, uses, and life-history are yet fully known to us. Our museums contain large numbers which botanists have not yet had time to describe and name. Even in our own country not a year passes without some additional plant being discovered ; as regards the less known regions of the earth not half the species have yet been collected. Among the Lichens and Fungi especially many problems of their life-history, some, indeed, of especial importance to man, still await solution.

Our knowledge of the fossil forms, moreover, falls far short even of that of existing species, which, on the other hand, they must have greatly exceeded in number. Every difference of form, structure, and colour has doubtless some cause and explanation, so that the field for research is really inexhaustible.

CHAPTER V

WOODS AND FIELDS

"By day or by night, summer or winter, beneath trees the heart feels nearer to that depth of life which the far sky means. The rest of spirit, found only in beauty, ideal and pure, comes there because the distance seems within touch of thought." JEFFERIES.

CHAPTER V

RURAL life, says Cicero, "is not delightful by reason of cornfields only and meadows, and vineyards and groves, but also for its gardens and orchards, for the feeding of cattle, the swarms of bees, and the variety of all kinds of flowers." Bacon considered that a garden is "the greatest refreshment to the spirits of man, without which buildings and palaces are but gross handyworks, and a man shall ever see, that when ages grow to civility and elegancy men come to build stately sooner than to garden finely, as if gardening were the greater perfection."

No doubt "the pleasure which we take in a garden is one of the most innocent delights in human life."[1] Elsewhere there may be scat-

---

[1] *The Spectator.*

tered flowers, or sheets of colour due to one or
two species, but in gardens one glory follows
another.   Here are brought together all the

> quaint enamelled eyes,
> That on the green turf sucked the honeyed showers,
> And purple all the ground with vernal flowers.
> Bring the rathe primrose that forsaken dies,
> The tufted crow-toe, and pale jessamine,
> The white pink and the pansy freaked with jet,
> The glowing violet,
> The musk rose, and the well attired woodbine,
> With cowslips wan that hang the pensive head,
> And every flower that sad embroidery wears.[1]

We cannot, happily we need not try to,
contrast or compare the beauty of gardens
with that of woods and fields.

And yet to the true lover of Nature wild
flowers have a charm which no garden can
equal. Cultivated plants are but a living
herbarium. They surpass, no doubt, the
dried specimens of a museum, but, lovely as
they are, they can be no more compared with
the natural vegetation of our woods and fields
than the captives in the Zoological Gardens
with the same wild species in their native
forests and mountains.

[1] Milton.

Often indeed, our woods and fields rival gardens even in the richness of colour. We have all seen meadows white with Narcissus, glowing with Buttercups, Cowslips, early purple Orchis, or Cuckoo Flowers; cornfields blazing with Poppies; woods carpeted with Bluebells, Anemones, Primroses, and Forget-me-nots; commons with the yellow Lady's Bedstraw, Harebells, and the sweet Thyme; marshy places with the yellow stars of the Bog Asphodel, the Sun-dew sparkling with diamonds, Ragged Robin, the beautifully fringed petals of the Buckbean, the lovely little Bog Pimpernel, or the feathery tufts of Cotton Grass; hedgerows with Hawthorn and Traveller's Joy, Wild Rose and Honeysuckle, while underneath are the curious leaves and orange fruit of the Lords and Ladies, the snowy stars of the Stitchwort, Succory, Yarrow, and several kinds of Violets; while all along the banks of streams are the tall red spikes of the Loosestrife, the Hemp Agrimony, Water Groundsel, Sedges, Bulrushes, Flowering Rush, Sweet Flag, etc.

Many other sweet names will also at once

occur to us — Snowdrops, Daffodils and Hearts-
ease, Lady's Mantles and Lady's Tresses,
Eyebright, Milkwort, Foxgloves, Herb Roberts,
Geraniums, and among rarer species, at least
in England, Columbines and Lilies.

But Nature does not provide delights for
the eye only. The other senses are not forgotten. A thousand sounds — many delightful in themselves, and all by association —
songs of birds, hum of insects, rustle of leaves,
ripple of water, seem to fill the air.

Flowers again are sweet, as well as lovely.
The scent of pine woods, which is said to
be very healthy, is certainly delicious, and
the effect of Woodland scenery is good for
the mind as well as for the body.

" Resting quietly under an ash tree, with
the scent of flowers, and the odour of green
buds and leaves, a ray of sunlight yonder
lighting up the lichen and the moss on the
oak trunk, a gentle air stirring in the branches
above, giving glimpses of fleecy clouds sailing
in the ether, there comes into the mind a feeling of intense joy in the simple fact of living." [1]

[1] Jefferies.

The wonderful phenomenon of phosphorescence is not a special gift to the animal kingdom. Henry O. Forbes describes a forest in Sumatra: "The stem of every tree blinked with a pale greenish-white light which undulated also across the surface of the ground like moonlight coming and going behind the clouds, from a minute thread-like fungus invisible in the day-time to the unassisted eye; and here and there thick dumpy mushrooms displayed a sharp, clear dome of light, whose intensity never varied or changed till the break of day; long phosphorescent caterpillars and centipedes crawled out of every corner, leaving a trail of light behind them, while fire-flies darted about above like a lower firmament."[1]

Woods and Forests were to our ancestors the special scenes of enchantment.

The great Ash tree Yggdrasil bound together Heaven, Earth, and Hell. Its top reached to Heaven, its branches covered the Earth, and the roots penetrated into Hell. The three Normas or Fates sat under it, spinning the thread of life.

[1] Forbes, *A Naturalist's Wanderings in the Eastern Archipelago.*

Of all the gods and goddesses of classical mythology or our own folk-lore, none were more fascinating than the Nature Spirits — Elves and Fairies, Neckans and Kelpies, Pixies and Ouphes, Mermaids, Undines, Water Spirits, and all the Elfin world

> Which have their haunts in dale and piny mountain,
> Or forests, by slow stream or tingling brook.

They come out, as we are told, especially on moonlight nights. But while evening thus clothes many a scene with poetry, forests are fairy land all day long.

Almost any wood contains many and many a spot well suited for Fairy feasts; where one might most expect to find Titania, resting, as once we are told,

> She lay upon a bank, the favourite haunt
> Of the Spring wind in its first sunshine hour,
> For the luxuriant strawberry blossoms spread
> Like a snow shower then, and violets
> Bowed down their purple vases of perfume
> About her pillow, — linked in a gay band
> Floated fantastic shapes; these were her guards,
> Her lithe and rainbow elves.

The fairies have disappeared, and, so far as

England is concerned, the larger forest animals have vanished almost as completely. The Elk and Bear, the Boar and Wolf have gone, the Stag has nearly disappeared, and but a scanty remnant of the original wild Cattle linger on at Chillingham. Still the woods teem with life; the Fox and Badger, Stoat and Weasel, Hare and Rabbit, and Hedgehog,

> The tawny squirrel vaulting through the boughs,
> Hawk, buzzard, jay, the mavis and the merle,[1]

the Owls and Nightjar, the Woodpecker, Nuthatch, Magpie, Doves, and a hundred more.

In early spring the woods are bright with the feathery catkins of the Willow, followed by the soft green of the Beech, the white or pink flowers of the Thorn, the pyramids of the Horse-chestnut, festoons of the Laburnum and Acacia, and the Oak slowly wakes from its winter sleep, while the Ash leaves long linger in their black buds.

Under foot is a carpet of flowers — Anemones, Cowslips, Primroses, Bluebells, and

---

[1] Tennyson.

the golden blossoms of the Broom, which, however, while Gorse and Heather continue in bloom for months, "blazes for a week or two, and is then completely extinguished, like a fire that has burnt itself out." [1]

In summer the tints grow darker, the birds are more numerous and full of life; the air teems with insects, with the busy murmur of bees and the idle hum of flies, while the cool of morning and evening, and the heat of the day, are all alike delicious.

As the year advances and the flowers wane, we have many beautiful fruits and berries, the red hips and haws of the wild roses, scarlet holly berries, crimson yew cups, the translucent berries of the Guelder Rose, hanging coral beads of the Black Bryony, feathery festoons of the Traveller's Joy, and others less conspicuous, but still exquisite in themselves — acorns, beech nuts, ash keys, and many more. It is really difficult to say which are most beautiful, the tender greens of spring or the rich tints of autumn, which glow so brightly in the sunshine.

[1] Hamerton.

Tropical fruits are even more striking. No one who has seen it can ever forget a grove of orange trees in full fruit; while the more we examine the more we find to admire; all perfectly and exquisitely finished "usque ad ungues," perfect inside and outside, for Nature

Does in the Pomegranate close
Jewels more rare than Ormus shows.[1]

· In winter the woods are comparatively bare and lifeless, even the Brambles and Woodbine, which straggle over the tangle of underwood being almost leafless.

Still even then they have a beauty and interest of their own; the mossy boles of the trees; the delicate tracery of the branches which can hardly be appreciated when they are covered with leaves; and under foot the beds of fallen leaves; while the evergreens seem brighter than in summer; the ruddy stems and rich green foliage of the Scotch Pines, and the dark spires of the Firs, seeming to acquire fresh beauty.

[1] Marvell.

Again in winter, though no doubt the living tenants of the woods are much less numerous, many of our birds being then far away in the dense African forests, on the other hand those which remain are much more easily visible. We can follow the birds from tree to tree, and the Squirrel from bough to bough.

It requires little imagination to regard trees as conscious beings, indeed it is almost an effort not to do so.

" The various action of trees rooting themselves in inhospitable rocks, stooping to look into ravines, hiding from the search of glacier winds, reaching forth to the rays of rare sunshine, crowding down together to drink at sweetest streams, climbing hand in hand among the difficult slopes, opening in sudden dances among the mossy knolls, gathering into companies at rest among the fragrant fields, gliding in grave procession over the heavenward ridges — nothing of this can be conceived among the unvexed and unvaried felicities of the lowland forest; while to all these direct sources of greater beauty are

added, first the power of redundance, the
mere quantity of foliage visible in the folds
and on the promontories of a single Alp
being greater than that of an entire lowland
landscape (unless a view from some Cathedral
tower) ; and to this charm of redundance, that
of clearer visibility — tree after tree being constantly shown in successive height, one behind
another, instead of the mere tops and flanks
of masses as in the plains ; and the forms of
multitudes of them continually defined against
the clear sky, near and above, or against
white clouds entangled among their branches,
instead of being confused in dimness of
distance." [1]

There is much that is interesting in the
relations of one species to another. Many
plants are parasitic upon others. The foliage
of the Beech is so thick that scarcely anything
will grow under it, except those spring plants,
such as the Anemone and the Wood Buttercup or Goldilocks, which flower early before
the Beech is in leaf.

There are other cases in which the reason

[1] Ruskin.

N

for the association of species is less evident. The Larch and the Arolla (Pinus Cembra) are close companions. They grow together in Siberia ; they do not occur in Scandinavia or Russia, but both reappear in certain Swiss valleys, especially in the cantons of Lucerne and Valais and the Engadine.

Another very remarkable case which has recently been observed is the relation existing between some of our forest trees and certain Fungi, the species of which have not yet been clearly ascertained. The root tips of the trees are as it were enclosed in a thin sheet of closely woven mycelium. It was at first supposed that the fungus was attacking the roots of the tree, but it is now considered that the tree and the fungus mutually benefit one another. The fungus collects nutriment from the soil, which passes into the tree and up to the leaves, where it is elaborated into sap, the greater part being utilized by the tree, but a portion reabsorbed by the fungus. There is reason to think that, in some cases at any rate, the mycelium is that of the Truffle.

The great tropical forests have a totally different character from ours. I reproduce here the plate from Kingsley's *At Last*. The trees strike all travellers by their magnificence, the luxuriance of their vegetation, and their great variety. Our forests contain comparatively few species, whereas in the tropics we are assured that it is far from common to see two of the same species near one another. But while in our forests the species are few, each tree has an independence and individuality of its own. In the tropics, on the contrary, they are interlaced and interwoven, so as to form one mass of vegetation; many of the trunks are almost concealed by an undergrowth of verdure, and intertwined by spiral stems of parasitic plants; from tree to tree hang an inextricable network of lianas, and it is often difficult to tell to which tree the fruits, flowers, and leaves really belong. The trunks run straight up to a great height without a branch, and then form a thick leafy canopy far overhead; a canopy so dense that even the blaze of the cloudless blue sky is subdued, one might almost say into a weird gloom, the effect of which is enhanced by the solemn silence. At first such a forest gives the impression of being more open than an English wood, but a few steps are sufficient to correct this error. There is a thick under- growth matted together by wiry creepers, and the intermediate space is traversed in all directions by lines and cords.

The English traveller misses sadly the sweet songs of our birds, which are replaced by the hoarse chatter of parrots. Now and then a succession of cries even harsher and more discordant tell of a troop of monkeys passing across from tree to tree among the higher branches, or lower sounds indicate to a practised ear the neighbourhood of an ape, a sloth, or some other of the few mammals which inhabit the great forests. Occasionally a large blue bee hums past, a brilliant butter- fly flashes across the path, or a humming-bird hangs in the air over a flower like, as St. Pierre says, an emerald set in coral, but " how weak it is to say that that exquisite little being, whirring and fluttering in the air, has a head of ruby, a throat of emerald, and wings of sapphire, as if any triumph of the jeweller's art could ever vie with that sparkling epitome of life and light." [1]

Sir Wyville Thomson graphically describes a morning in a Brazilian forest : —

" The night was almost absolutely silent, only now and then a peculiarly shrill cry of some night bird reached us from the woods. As we got into the skirt of the forest the morning broke, but the *réveil* in a Brazilian forest is wonderfully different from the slow creeping on of the dawn of a summer morning at home, to the music of the thrushes answering one another's full rich notes from neighbouring thorn-trees. Suddenly a yellow light spreads upwards in the east, the stars quickly fade, and the dark fringes of the forest and the tall palms show out black against the yellow sky, and almost before one has time to observe the change the sun has risen straight and fierce, and the whole landscape is bathed in the full light of day. But the morning is yet for another hour cool and fresh, and the scene is indescribably beautiful. The woods,

[1] Thomson, *Voyage of the Challenger.*

so absolutely silent and still before, break at once into noise and movement. Flocks of toucans flutter and scream on the tops of the highest forest trees hopelessly out of shot, the ear is pierced by the strange wild screeches of a little band of macaws which fly past you like the wrapped-up ghosts of the birds on some gaudy old brocade." [1]

Mr. Darwin tells us that nothing can be better than the description of tropical forests given by Bates.

"The leafy crowns of the trees, scarcely two of which could be seen together of the same kind, were now far away above us, in another world as it were. We could only see at times, where there was a break above, the tracery of the foliage against the clear blue sky. Sometimes the leaves were palmate, or of the shape of large outstretched hands; at others finely cut or feathery like the leaves of Mimosæ. Below, the tree trunks were everywhere linked together by sipos; the woody flexible stems of climbing and creeping trees, whose foliage is far away above, mingled with

---

[1] Thomson, *Voyage of the Challenger.*

that of the taller independent trees. Some
were twisted in strands like cables, others had
thick stems contorted in every variety of shape,
entwining snake-like round the tree trunks or
forming gigantic loops and coils among the
larger branches; others, again, were of zigzag
shape, or indented like the steps of a staircase,
sweeping from the ground to a giddy height."

The reckless and wanton destruction of
forests has ruined some of the richest countries
on earth. Syria and Asia Minor, Palestine
and the north of Africa were once far more
populous than they are at present. They were
once lands "flowing with milk and honey,"
according to the picturesque language of the
Bible, but are now in many places reduced to
dust and ashes. Why is there this melancholy
change? Why have deserts replaced cities?
It is mainly owing to the ruthless destruction
of the trees, which has involved that of
nations. Even nearer home a similar process
may be witnessed. Two French departments
— the Hautes- and Basses-Alpes — are being
gradually reduced to ruin by the destruction
of the forests. Cultivation is diminishing,

vineyards are being washed away, the towns are threatened, the population is dwindling, and unless something is done the country will be reduced to a desert; until, when it has been released from the destructive presence of man, Nature reproduces a covering of vegetable soil, restores the vegetation, creates the forests anew, and once again fits these regions for the habitation of man.

In another part of France we have an illustration of the opposite process.

The region of the Landes, which fifty years ago was one of the poorest and most miserable in France, has now been made one of the most prosperous owing to the planting of Pines. The increased value is estimated at no less than 1,000,000,000 francs. Where there were fifty years ago only a few thousand poor and unhealthy shepherds whose flocks pastured on the scanty herbage, there are now sawmills, charcoal kilns, and turpentine works, interspersed with thriving villages and fertile agricultural lands.

In our own country, though woodlands are perhaps on the increase, true forest scenery is gradually disappearing. This is, I suppose, unavoidable, but it is a matter of regret. Forests have so many charms of their own. They give a delightful impression of space and of abundance.

The extravagance is sublime. Trees, as Jefferies says, " throw away handfuls of flower; and in the meadows the careless, spendthrift ways of grass and flower and all things are not to be expressed. Seeds by the hundred million float with absolute indifference on the air. The oak has a hundred thousand more leaves than necessary, and never hides a single acorn. Nothing utilitarian — everything on a scale of splendid waste. Such noble, broadcast, open-armed waste is delicious to behold. Never was there such a lying proverb as ' Enough is as good as a feast.' Give me the feast; give me squandered millions of seeds, luxurious carpets of petals, green mountains of oak-leaves. The greater the waste the greater the enjoyment — the nearer the approach to real life."

It is of course impossible here to give any idea of the complexity of structure of our

forest trees. A slice across the stem of a
tree shows many different tissues with more or
less technical names, bark and cambium, med-
ullary rays, pith, and more or less specialised
tissue; air-vessels, punctate vessels, woody
fibres, liber fibres, scalariform vessels, and
other more or less specialised tissues.

Let us take a single leaf. The name is
synonymous with anything very thin, so that
we might well fancy that a leaf would consist
of only one or two layers of cells. Far from
it, the leaf is a highly complex structure. On
the upper surface are a certain number of
scattered hairs, while in the bud these are
often numerous, long, silky, and serve to
protect the young leaf, but the greater number
fall off soon after the leaf expands. The hairs
are seated on a layer of flattened cells — the
skin or epidermis. Below this are one or
more layers of " palisade cells," the function
of which seems to be to regulate the quantity
of light entering the leaf. Under these again
is the " parenchyme," several layers of more or
less rounded cells, leaving air spaces and passages between them. From place to place in

the parenchyme run "fibro-vascular bundles," forming a sort of skeleton to the leaf, and comprising air-vessels on the upper side, rayed or dotted vessels with woody fibre below, and vessels of various kinds. The under surface of the leaf is formed by another layer of flattened cells, supporting generally more or less hairs, and some of them specially modified so as to leave minute openings or "stomata" leading into the air passages. These stomata are so small that there are millions on a single leaf, and on plants growing in dry countries, such as the Evergreen Oak Oleander, etc., they are sunk in pits, and further protected by tufts of hair.

The cells of the leaf again are themselves complex. They consist of a cell wall perforated by extremely minute orifices, of protoplasm, cell fluid, and numerous granules of "Chlorophyll," which give the leaf its green colour.

While these are, stated very briefly, the essential parts of a leaf, the details differ in every species, while in the same species and even in the same plant. the leaves present

minor differences according to the situation
in which they grow.

Since, then, there is so much complex
structure in a single leaf, what must it be in a
whole plant? There is a giant seaweed (Mac-
rocystis), which has been known to reach a
length of 1000 feet, as also do some of the
lianas of tropical forests. These, however,
attain no great bulk, and the most gigantic
specimens of the vegetable kingdom yet
known are the Wellingtonia (Sequoia) gigan-
tea, which grows to a height of 450 feet, and
the Blue Gum (Eucalyptus) even to 480.

One is apt to look on animal structure as
more delicate, and of a higher order, than
that of plants. And so no doubt it is. Yet
an animal, even man himself, will recover
from a wound or an operation more rapidly
and more perfectly than a tree.[1]

Trees again derive a special interest from
the venerable age they attain. In some cases,
no doubt, the age is more or less mythical, as,
for instance, the Olive of Minerva at Athens,
the Oaks mentioned by Pliny, "which were

---

[1] Sir J. Paget, *On the Pathology of Plants.*

thought coeval with the world itself," the
Fig tree, "under which the wolf suckled the
founder of Rome and his brother, lasting (as
Tacitus calculated) 840 years, putting out
new shoots, and presaging the translation of
that empire from the Cæsarian line, happening in Nero's reign."[1] But in other cases the
estimates rest on a surer foundation, and it
cannot be doubted that there are trees still
living which were already of considerable size
at the time of the Conquest. The Soma
Cypress of Lombardy, which is 120 feet high
and 23 in circumference, is calculated to go
back to forty years before the birth of Christ.
Francis the First is said to have driven his
sword into it in despair after the battle of
Padua, and Napoleon altered his road over the
Simplon so as to spare it.

Ferdinand and Isabella in 1476 swore
to maintain the privileges of the Biscayans
under the old Oak of Guernica. In the
Ardennes an Oak cut down in 1824 contained a funeral urn and some Samnite
coins. A writer at the time drew the conclu-

[1] Evelyn's *Sylva*.

sion that it must have been already a large
tree when Rome was founded, and though the
facts do not warrant this conclusion, the tree
did, no doubt, go back to Pagan times. The
great Yew of Fountains Abbey is said to have
sheltered the monks when the abbey was re-
built in 1133, and is estimated at an age of
1300 years; that at Brabourne in Kent at
3000. De Candolle gives the following as the
ages attainable : —

The Ivy	450 years
Larch	570 "
Plane	750 "
Cedar of Lebanon	800 "
Lime	1100 "
Oak	1500 "
Taxodium distichum	4000 to 6000
Baobab	6000 years

Nowhere is woodland scenery more beautiful than where it passes gradually into the
open country. The separate trees, having
more room both for their roots and branches.
are finer, and can be better seen, while, when
they are close together, " one cannot see the
wood for the trees." The vistas which open
out are full of mystery and of promise.

and tempt us gradually out into the green fields.

What pleasant memories these very words recall, games in the hay as children, and sunny summer days throughout life.

"Consider," says Ruskin,[1] " what we owe to the meadow grass, to the covering of the dark ground by that glorious enamel, by the companies of those soft countless and peaceful spears. The fields! Follow but forth for a little time the thought of all that we ought to recognise in those words. All spring and summer is in them — the walks by silent scented paths, the rests in noonday heat, the joy of herds and flocks, the power of all shepherd life and meditation, the life of sunlight upon the world, falling in emerald streaks, and soft blue shadows, where else it would have struck on the dark mould or scorching dust, pastures beside the pacing brooks, soft banks and knolls of lowly hills, thymy slopes of down overlooked by the blue line of lifted sea, crisp lawns all dim with early dew, or smooth in evening warmth of

[1] *Modern Painters.*

barred sunshine, dinted by happy feet. and softening in their fall the sound of loving voices.

.   .   .   .   .   .

" Go out, in the spring time, among the meadows that slope from the shores of the Swiss lakes to the roots of their lower mountains.   There, mingled with the taller gentians and the white narcissus, the grass grows deep and free, and as you follow the winding mountain paths, beneath arching boughs all veiled and dim with blossom, — paths. that for ever droop and rise over the green banks and mounds sweeping down in scented undulation. steep to the blue water, studded here and there with new mown heaps, filling all the air with fainter sweetness, — look up towards the higher hills. where the waves of everlasting green roll silently into their long inlets among the shadows of the pines ; and we may. perhaps, at last know the meaning of those quiet words of the 147th Psalm, ' He maketh the grass to grow upon the mountains.' "

" On fine days," he tells us again in his *Autobiography,* " when the grass was dry, I

used to lie down on it, and draw the blades
as they grew, with the ground herbage of
buttercup or hawkweed mixed among them,
until every square foot of meadow, or mossy
bank, became an infinite picture and possession to me, and the grace and adjustment to
each other of growing leaves, a subject of
more curious interest to me than the composition of any painter's masterpieces."

In the passage above quoted, Ruskin alludes
especially to Swiss meadows. They are especially remarkable in the beauty and variety of
flowers. In our fields the herbage is mainly
grass, and if it often happens that they glow
with Buttercups or are white with Ox-eye-
daisies, these are but unwelcome intruders
and add nothing to the value of the hay.
Swiss meadows, on the contrary, are sweet
and lovely with wild Geraniums, Harebells,
Bluebells, Pink Restharrow, Yellow Lady's
Bedstraw, Chervil, Eyebright, Red and White
Silenes, Geraniums, Gentians, and many other
flowers which have no familiar English names ;
all adding not only to the beauty and sweetness
of the meadows, but forming a valuable part

of the crop itself.[1]   On the other hand " turf "
is peculiarly English, and no turf is more delightful than that of our Downs — delightful
to ride on, to sit on, or to walk on.   The turf
indeed feels so springy under our feet that
walking on it seems scarcely an exertion : one
could almost fancy that the Downs themselves
were still rising, even higher, into the air.

The herbage of the Downs is close rather
than short, hillocks of sweet thyme, tufts of
golden Potentilla, of Milkwort — blue, pink,
and white — of sweet grass and Harebells :
here and there pink with Heather, or golden
with Furze or Broom, while over all are the
fresh air and sunshine, sweet scents, and the
hum of bees.   And if the Downs seem full of
life and sunshine, their broad shoulders are
types of kindly strength, they give also an
impression of power and antiquity, while every
now and then we come across a tumulus, or a
group of great grey stones, the burial place of
some ancient hero, or a sacred temple of our
pagan forefathers.

[1] M. Correvon informs me that the Gruyère cheese is supposed
to owe its peculiar flavour to the alpine Alchemilla, which is now
on that account often purposely sown elsewhere.

On the Downs indeed things change slowly,
and in parts of Sussex the strong slow oxen
still draw the waggons laden with warm hay
or golden wheat sheaves, or drag the wooden
plough along the slopes of the Downs, just as
they did a thousand years ago.

I love the open Down most, but without
hedges England would not be England.
Hedges are everywhere full of beauty and
interest, and nowhere more so than at the
foot of the Downs, when they are in great
part composed of wild Guelder Roses and rich
dark Yews, decked with festoons of Traveller's Joy, the wild Bryonies, and garlands of
Wild Roses covered with thousands of white
or delicate pink flowers, each with a centre of
gold.

At the foot of the Downs spring clear sparkling streams; rain from heaven purified still
further by being filtered through a thousand
feet of chalk; fringed with purple Loosestrife
and Willowherb, starred with white Water
Ranunculuses, or rich Watercress, while every
now and then a brown water rat rustles in
the grasses at the edge, and splashes into

the water, or a pink speckled trout glides out of sight.

In many of our midland and northern counties most of the meadows lie in parallel undulations or "rigs." These are generally about a furlong (220 yards) in length, and either one or two poles (5½ or 11 yards) in breadth. They seldom run straight, but tend to curve towards the left. At each end of the field a high bank, locally called a balk, often 3 or 4 feet high, runs at right angles to the rigs. In small fields there are generally eight, but sometimes ten, of these rigs, which make in the one case 4, in the other 5 acres. These curious characters carry us back to the old tenures, and archaic cultivation of land, and to a period when the fields were not in pasture, but were arable.

They also explain our curious system of land measurement. The "acre" is the amount which a team of oxen were supposed to plough in a day. It corresponds to the German "morgen" and the French "journée." The furlong or long "furrow" is the distance which a team of oxen can plough conven-

iently without stopping to rest. Oxen, as we
know, were driven not with a whip, but with
a goad or pole, the most convenient length for
which was 16½ feet, and the ancient ploughman used his " pole " or " perch " by placing
it at right angles to his first furrow, thus
measuring the amount he had to plough.
Hence our " pole " or " perch " of 16½ feet,
which at first sight seems a very singular
unit to have selected. This width is also convenient both for turning the plough, and also
for sowing. Hence the most convenient unit
of land for arable purposes was a furlong in
length and a perch or pole in width.

The team generally consisted of eight oxen.
Few peasants, however, possessed a whole
team, several generally joining together, and
dividing the produce. Hence the number of
" rigs," one for each ox. We often, however,
find ten instead of eight ; one being for the
parson's tithe, the other tenth going to the
ploughman.

When eight oxen were employed the goad
would not of course reach the leaders, which
were guided by a man who walked on the

near side. On arriving at the end of each
furrow he turned them round, and as it was
easier to pull than to push them, this gradually gave the furrow a turn towards the left,
thus accounting for the slight curvature.
Lastly, while the oxen rested on arriving at
the end of the furrow, the ploughmen scraped
off the earth which had accumulated on the
coulter and ploughshare, and the accumulation
of these scrapings gradually formed the balk.

It is fascinating thus to trace indications
of old customs and modes of life, but it would
carry us away from the present subject.

Even though the Swiss meadows may offer
a greater variety, our English fields are yet
rich in flowers: yellow with Cowslips and
Primroses, pink with Cuckoo flowers and
purple with Orchis, while, however, unwelcome to the eye of the farmer,

> the rich Buttercup
> Its tiny polished urn holds up,
> Filled with ripe summer to the edge,[1]

turning many a meadow into a veritable field
of the cloth of gold, and there are few prettier

[1] J. R. Lowell.

sights in nature than an English hay field on
a summer evening, with a copse perhaps at
one side and a brook on the other; men with
forks tossing the hay in the air to dry;
women with wooden rakes arranging it in
swathes ready for the great four-horse wag-
gon, or collecting it in cocks for the night;
while some way off the mowers are still at
work, and we hear from time to time the
pleasant sound of the whetting of the scythe.
All are working with a will lest rain should
come and their labour be thrown away. This
too often happens. But though we often complain of our English climate, it is yet, take
it all in all, one of the best in the world,
being comparatively free from extremes either
of heat or cold, drought or deluge. To the
happy mixture of sunshine and of rain we
owe the greenness of our fields,

> sparkling with dewdrops
> Indwelt with little angels of the Sun, [1]

lit and

> warmed by golden sunshine
> And fed by silver rain,

which now and again sprinkles the whole earth
with diamonds.

[1] Hamerton.

CHAPTER VI

MOUNTAINS

Mountains " seem to have been built for the human race, as at once their schools and cathedrals; full of treasures of illuminated manuscript for the scholar, kindly in simple lessons for the worker, quiet in pale cloisters for the thinker, glorious in holiness for the worshipper. They are great cathedrals of the earth, with their gates of rock. pavements of cloud, choirs of stream and stone, altars of snow, and vaults of purple traversed by the continual stars."— RUSKIN.

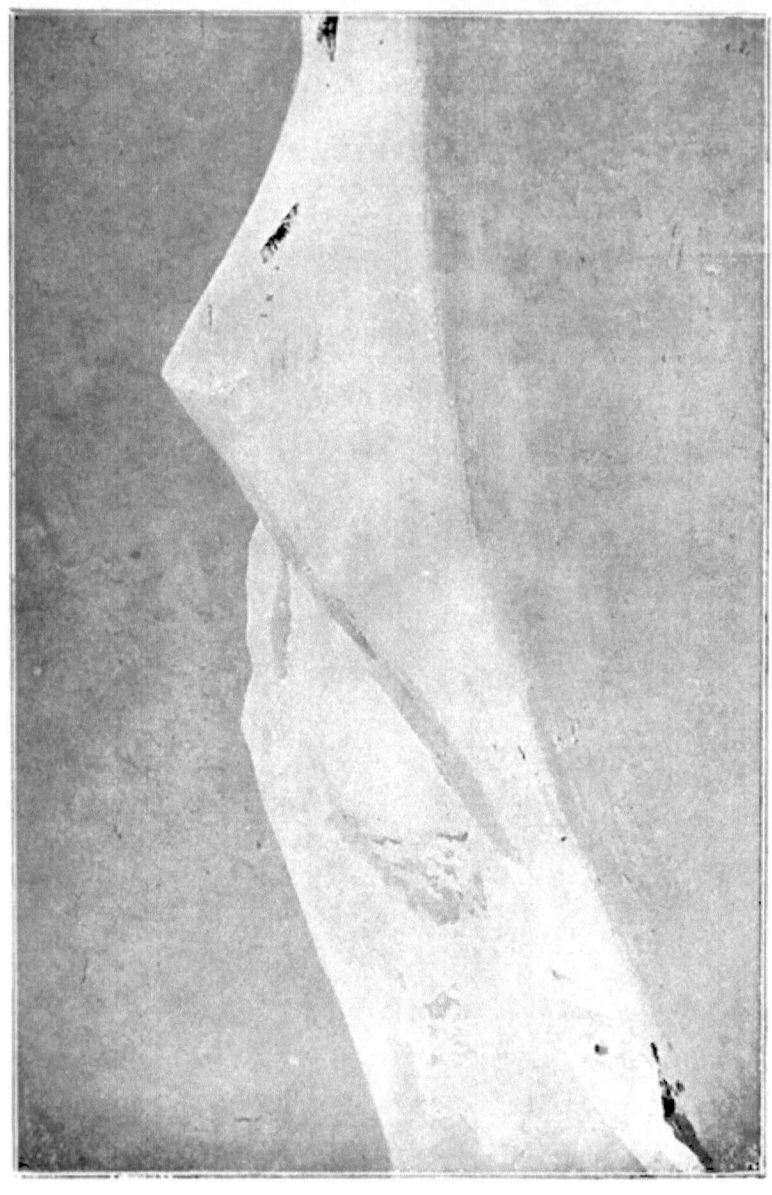

SUMMIT OF MONT BLANC.

CHAPTER VI

MOUNTAINS

THE Alps are to many of us an inexhaustible source of joy and peace, of health, and even of life. We have gone to them jaded and worn, feeling, perhaps without any external cause, anxious and out of spirits, and have returned full of health, strength, and energy. Among the mountains Nature herself seems freer and happier, brighter and purer, than elsewhere. The rush of the rivers, and the repose of the lakes, the pure snowfields and majestic glaciers, the fresh air, the mysterious summits of the mountains, the blue haze of the distance, the morning tints and the evening glow, the beauty of the sky and the grandeur of the storm, have all refreshed and delighted us time after time, and their memories can never fade away.

Even now as I write comes back to me the bright vision of an Alpine valley — blue sky above, glittering snow, bare grey or rich red rock, dark pines here and there, mixed with bright green larches, then patches of smooth alp, with clumps of birch and beech, and dotted with brown châlets; then below them rock again, and wood, but this time with more deciduous trees; and then the valley itself, with emerald meadows, interspersed with alder copses, threaded together by a silver stream; and I almost fancy I can hear the tinkling of distant cowbells coming down from the alp, and the delicious murmur of the rushing water. The endless variety, the sense of repose and yet of power, the dignity of age, the energy of youth, the play of colour, the beauty of form, the mystery of their origin, all combine to invest mountains with a solemn beauty.

I feel with Ruskin that " mountains are the beginning and the end of all natural scenery; in them, and in the forms of inferior landscape that lead to them, my affections are wholly bound up; and though I can look with happy admiration at the lowland flowers, and woods, and open skies, the happiness is tranquil and cold, like that of examining detached flowers in a conservatory, or reading a pleasant book." And of all mountain views which he has seen, the finest he considers is that from the Montanvert : " I have climbed much and wandered much in the heart of the high Alps, but I have never yet seen anything which equalled the view from the cabin of the Montanvert."

It is no mere fancy that among mountains the flowers are peculiarly large and brilliant in colour. Not only are there many beautiful species which are peculiar to mountains, — alpine Gentians, yellow, blue, and purple ; alpine Rhododendrons, alpine Primroses and Cowslips, alpine Lychnis, Columbine, Monkshood, Anemones, Narcissus, Campanulas, Soldanellas, and a thousand others less familiar to us, — but it is well established that even within the limits of the same species those living up in the mountains have larger and brighter flowers than their sisters elsewhere.

Various alpine species belonging to quite distinct families form close moss-like cushions, gemmed with star-like flowers, or covered completely with a carpet of blossom. On the lower mountain slopes and in alpine valleys trees seem to flourish with peculiar luxuriance. Pines and Firs and Larches above; then, as we descend, Beeches and magnificent Chestnuts, which seem to rejoice in the sweet, fresh air and the pure mountain streams.

To any one accustomed to the rich bird life of English woods and hedgerows, it must be admitted that Swiss woods and Alps seem rather lonely and deserted. Still the Hawk, or even Eagle, soaring high up in the air, the weird cry of the Marmot, and the knowledge that, even if one cannot see Chamois, they may all the time be looking down on us, give the Alps, from this point of view also, a special interest of their own.

Another great charm of mountain districts is the richness of colour. "Consider,[1] first. the difference produced in the whole tone of landscape colour by the introductions of purple. violet. and deep ultra-marine blue which we owe to mountains. In an ordinary lowland landscape we have the blue of the sky; the

---

[1] Ruskin.

green of the grass, which I will suppose (and
this is an unnecessary concession to the lowlands) entirely fresh and bright; the green of
trees; and certain elements of purple, far
more rich and beautiful than we generally
should think, in their bark and shadows (bare
hedges and thickets, or tops of trees, in subdued afternoon sunshine, are nearly perfect
purple and of an exquisite tone), as well as in
ploughed fields, and dark ground in general.
But among mountains, in addition to all this,
large unbroken spaces of pure violet and
purple are introduced in their distances; and
even near, by films of cloud passing over the
darkness of ravines or forests, blues are produced of the most subtle tenderness; these
azures and purples passing into rose colour of
otherwise wholly unattainable delicacy among
the upper summits, the blue of the sky being
at the same time purer and deeper than in the
plains. Nay, in some sense, a person who
has never seen the rose colour of the rays of
dawn crossing a blue mountain twelve or
fifteen miles away can hardly be said to know
what tenderness in colour means at all; bright

tenderness he may, indeed, see in the sky or
in a flower, but this grave tenderness of the
far-away hill-purples he cannot conceive."

"I do not know," he says elsewhere, "any
district possessing a more pure or uninter-
rupted fulness of mountain character (and
that of the highest order), or which appears to
have been less disturbed by foreign agencies,
than that which borders the course of the
Trient between Valorsine and Martigny.   The
paths which lead to it, out of the valley of the
Rhone, rising at first in steep circles among
the walnut trees, like winding stairs among
the pillars of a Gothic tower, retire over the
shoulders of the hills into a valley almost
unknown, but thickly inhabited by an industrious and patient population.   Along the
ridges of the rocks, smoothed by old glaciers,
into long, dark, billowy swellings, like the
backs of plunging dolphins, the peasant
watches the slow colouring of the tufts of moss
and roots of herb, which, little by little, gather
a feeble soil over the iron substance ; then,
supporting the narrow strip of clinging ground
with a few stones, he subdues it to the spade,

and in a year or two a little crest of corn is seen waving upon the rocky casque."

Tyndall, speaking of the scene from the summit of the Little Scheideck,[1] says: "The upper air exhibited a commotion which we did not experience; clouds were wildly driven against the flanks of the Eiger, the Jungfrau thundered behind, while in front of us a magnificent rainbow, fixing one of its arms in the valley of Grindelwald, and, throwing the other right over the crown of the Wetterhorn, clasped the mountain in its embrace. Through jagged apertures in the clouds floods of golden light were poured down the sides of the mountain. On the slopes were innumerable châlets, glistening in the sunbeams, herds browsing peacefully and shaking their mellow bells; while the blackness of the pine trees, crowded into woods, or scattered in pleasant clusters over alp and valley, contrasted forcibly with the lively green of the fields."

Few men had more experience of mountains than Mr. Whymper, and from him, I will quote one remarkable passage de-

---

[1] *The Glaciers of the Alps.*

scribing the view from the summit of the
Matterhorn just before the terrible catastrophe
which overshadows the memory of his first
ascent.

" The day was one of those superlatively
calm and clear ones which usually precede
bad weather. The atmosphere was perfectly
still and free from all clouds or vapours.
Mountains fifty, nay, a hundred miles off
looked sharp and near. All their details —
ridge and crag, snow and glacier — stood out
with faultless definition. Pleasant thoughts
of happy days in bygone years came up
unbidden as we recognised the old familiar
forms. All were revealed, not one of the
principal peaks of the Alps was hidden. I see
them clearly now, the great inner circle of
giants, backed by the ranges, chains, and
massifs. . . . Ten thousand feet beneath us
were the green fields of Zermatt, dotted with
châlets, from which blue smoke rose lazily.
Eight thousand feet below, on the other side,
were the pastures of Breuil. There were black
and gloomy forests; bright and cheerful
meadows, bounding waterfalls and tranquil

lakes, fertile lands and savage wastes, sunny
plains and frigid plateaux. There were the
most rugged forms and the most graceful
outlines, bold perpendicular cliffs and gentle
undulating slopes; rocky mountains and
snowy mountains, sombre and solemn, or
glittering and white, with walls, turrets, pinnacles, pyramids, domes, cones, and spires!
There was every combination that the world
can give, and every contrast that the heart
could desire."

These were summer scenes, but the
Autumn and Winter again have a grandeur
and beauty of their own.

" Autumn is dark on the mountains; grey
mist rests on the hills. The whirlwind is
heard on the heath. Dark rolls the river
through the narrow plain. The leaves twirl
round with the wind, and strew the grave of
the dead." [1]

Even bad weather often but enhances the
beauty and grandeur of mountains. When
the lower parts are hidden, and the peaks
stand out above the clouds, they look much

[1] Ossian.

loftier than if the whole mountain side is
visible. The gloom lends a weirdness and
mystery to the scene, while the flying clouds
give it additional variety.

Rain, moreover, adds vividness to the
colouring. The leaves and grass become a
brighter green, "every sunburnt rock glows
into an agate," and when fine weather returns
the new snow gives intense brilliance, and
invests the woods especially with the beauty
of Fairyland. How often in alpine districts
does one long "for the wings of a dove," more
thoroughly to enjoy and more completely to
explore, the mysteries and recesses of the
mountains. The mind, however, can go, even
if the body must remain behind.

Each hour of the day has a beauty of its
own. The mornings and evenings again glow
with different and even richer tints.

In mountain districts the cloud effects are
brighter and more varied than in flatter
regions. The morning and evening tints are
seen to the greatest advantage, and clouds
floating high in the heavens sometimes glitter
with the most exquisite iridescent hues

> that blush and glow
> Like angels' wings.[1]

On low ground one may be in the clouds, but not above them. But as we look down from mountains and see the clouds floating far below us, we almost seem as if we were looking down on earth from one of the heavenly bodies.

Not even in the Alps is there anything more beautiful than the " after glow " which lights up the snow and ice with a rosy tint for some time after the sun has set. Long after the lower slopes are already in the shade, the summit of Mont Blanc for instance is transfigured by the light of the setting sun glowing on the snow. It seems almost like a light from another world, and vanishes as suddenly and mysteriously as it came.

As we look up from the valleys the mountain peaks seem like separate pinnacles projecting far above the general level. This, however, is a very erroneous impression, and when we examine the view from the top of any of the higher mountains, or even from

[1] Bullar, *Azores.*

one of very moderate elevation, if well placed, such say as the well-known Piz Languard, we see that in many cases they must have once formed a dome, or even a table land, out of which the valleys have been carved. Many mountain chains were originally at least twice as high as they are now, and the highest peaks are those which have suffered least from the wear and tear of time.

We used to speak of the everlasting hills, and are only beginning to realise the vast and many changes which our earth has undergone.

> There rolls the deep where grew the tree.
>   O earth, what changes hast thou seen!
>   There where the long street roars, hath been
> The stillness of the central sea.
>
> The hills are shadows, and they flow
>   From form to form, and nothing stands;
>   They melt like mist, the solid lands,
> Like clouds they shape themselves and go.[1]

THE ORIGIN OF MOUNTAINS

Geography moreover acquires a new interest when we once realise that mountains

[1] Tennyson.

are no mere accidents, but that for every mountain chain, for every peak and valley, there is a cause and an explanation.

The origin of Mountains is a question of much interest. The building up of Volcanoes is even now going on before our eyes. Some others, the Dolomites for instance, have been regarded by Richthofen and other geologists as ancient coral islands. The long lines of escarpment which often stretch for miles across country, are now ascertained, mainly through the researches of Whitaker, to be due to the differential action of aerial causes. The general origin of mountain chains, however, was at first naturally enough attributed to direct upward pressure from below. To attribute them in any way to subsidence seems almost a paradox, and yet it appears to be now well established that the general cause is lateral compression, due to contraction of the underlying mass. The earth, we know, has been gradually cooling, and as it contracted in doing so, the strata of the crust would necessarily be thrown into folds. When an apple dries and shrivels in winter, the surface becomes covered

with ridges.   Or again, if we place some sheets
of paper between two weights on a table. and
then bring the weights nearer together, the
paper will be crumpled up.

In the same way let us take a section of
the earth's surface AB (Fig. 17), and suppose
that, by the gradual cooling and consequent
contraction of the mass, AB sinks to A′B′,

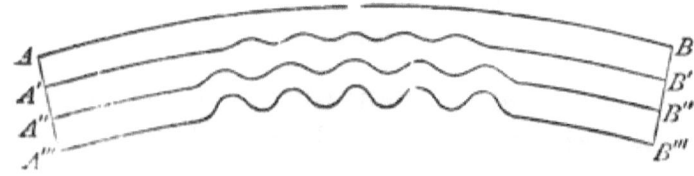

Fig. 17.—Adapted from Ball's paper " On the Formation of Alpine Valleys
and Lakes," *Lond. and Ed. Phil. Mag.* 1863, p. 96.

then to A″B″, and finally to A‴B‴.   Of
course if the cooling of the surface and of the
deeper portion were the same, then the strata
between A and B would themselves contract.
and might consequently still form a regular
curve between A‴ and B‴.   As a matter of
fact, however. the strata at the surface of our
globe have long since approached a constant
temperature.   Under these circumstances
there would be no contraction of the strata
between A and B corresponding to that of

those in the interior, and consequently they could not lie flat between A''' and B''', but must be thrown into folds, commencing along any line of least resistance. Sometimes indeed the strata are completely inverted, as in Fig. 19, and in other cases they have been squeezed for miles out of their original position. This explanation was first, I believe, suggested by Steno. It has been recently developed by Ball and Suess, and especially by Heim. In this manner it is probable that most mountain chains originated.[1]

The structure of mountain districts confirms this theoretical explanation. It is obvious of course that when strata are thrown into folds, they will, if strained too much, give way at the summit of the fold. Before doing so, however, they are stretched and consequently loosened, while on the other hand the strata at the bottom of the fold are compressed: the former, therefore, are rendered more susceptible of disintegration, the latter on the contrary acquire greater powers of resistance.

---

[1] See especially Heim's great work, *Unt. ü. d. Mechanismus der Gebirgsbildung.*

Hence denudation will act with more effect on the upper than on the lower portion of the folds, and if continued long enough, so that, as shown in the above diagram, the dotted portion is removed, we find the original hill tops replaced by valleys, and the original valleys forming the hill tops.   Every visitor to Switzerland must have noticed hills where the strata lie as shown in parts of Fig. 18, and where it is obvious that strata corresponding to those in dots must have been originally present.

In the Jura, for instance, a glance at any good map of the district will show a succession of ridges running parallel to one another in a slightly curved line from S.W. to N.E. That these ridges are due to folds of the earth's surface is clear from the following figure in Jaccard's work on the Geology of the Jura, showing a section from Brenets due south to Neuchâtel by Le Locle.   These folds are comparatively slight and the hills of no great height.   Further south, however, the strata are much more violently dislocated and compressed together.   The Mont Salève is the remnant of one of these ridges.

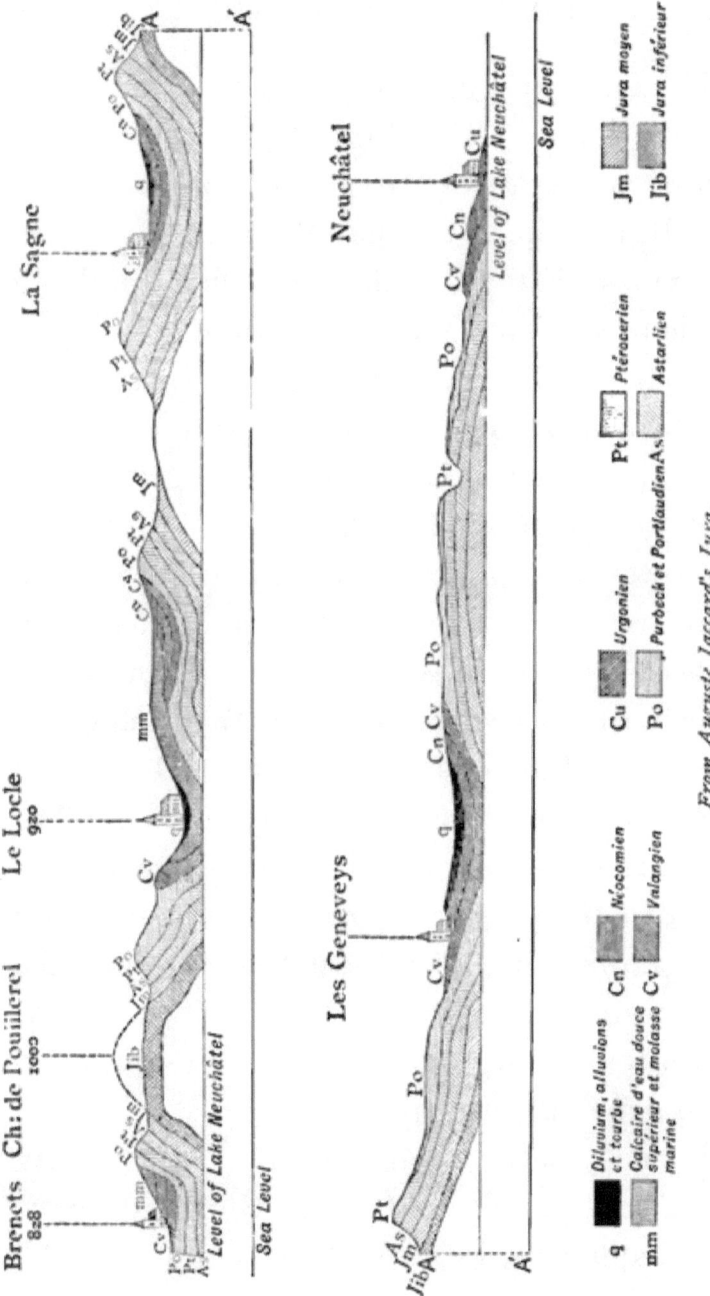

Fig. 18. — Section across the Jura from Brenets to Neuchâtel.

From Auguste Jaccard's Jura.

In the Alps the contortions are much greater than in the Jura. Fig. 19 shows a section after Heim, from the Spitzen across the Brunnialp, and the Maderanerthal. It is obvious that the valleys are due mainly to erosion, that the Maderaner valley has been cut out of the crystalline rocks *s*, and was once covered by the Jurassic strata *j*. which must have formerly passed in a great arch over what is now the valley.

However improbable it may seem that so great an amount of rock should have disappeared, evidence is conclusive. Ramsay has shown that in some parts of Wales not less than 29,000 feet have been removed. while there is strong reason for the belief that in Switzerland an amount has been carried away equal to the present height of the mountains; though of course it does not follow that the Alps were once twice as high as they are at present. because elevation and erosion must have gone on contemporaneously.

It has been calculated that the strata between Bâle and the St. Gotthard have been compressed from 202 miles to 130

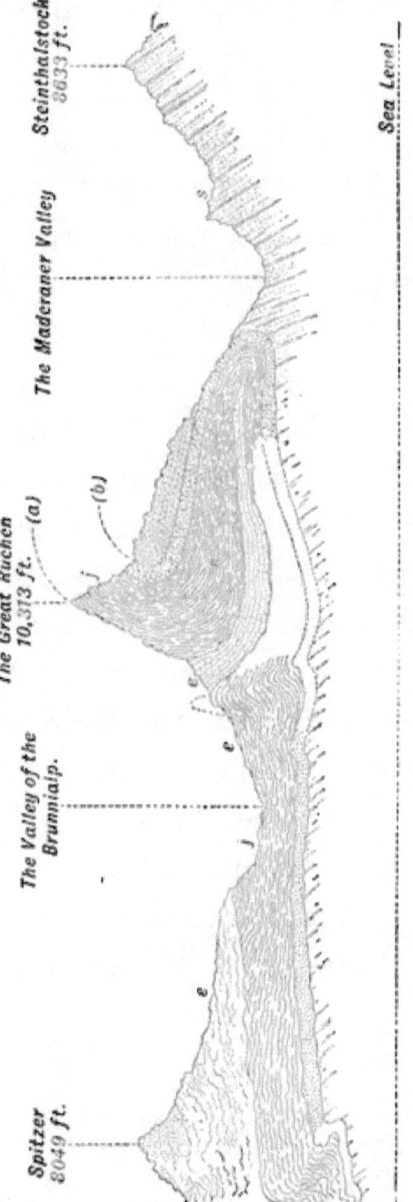

Fig. 19. — *e*, Eocene strata; *j*, Jurassic; *s*, Crystalline rocks.

miles, the Ardennes from 50 to 25 miles, and the Appalachians from 153 miles to 65! Prof. Gumbel has recently expressed the opinion that the main force to which the elevation of the Alps was due acted along the main axis of elevation. Exactly the opposite inference would seem really to follow from the facts. If the centre of force were along the axis of elevation, the result would, as Suess and Heim have pointed out, be to extend, not to compress, the strata; and the folds would remain quite unaccounted for. The suggestion of compression is on the contrary consistent with the main features of Swiss geography. The principal axis follows a curved line from the Maritime Alps towards the north-east by Mont Blanc and Monte Rosa and St. Gotthard to the mountains overlooking the Engadine. The geological strata follow the same direction. North of a line running through Chambery, Yverdun, Neuchâtel, Solothurn, and Olten to Waldshut on the Rhine are Jurassic strata: between that line and a second nearly parallel and running through Annecy, Vevey, Lucerne, Wesen,

Appenzell, and Bregenz on the Lake of Constance, is the lowland occupied by later Tertiary strata; between this second line and another passing through Albertville, St. Maurice, Lenk, Meiringen, and Altdorf lies a more or less broken band of older Tertiary strata; south of which are a Cretaceous zone, one of Jurassic age, then a band of crystalline rocks. while the central core, so to say, of the Alps, as for instance at St. Gotthard, consists mainly of gneiss or granite. The sedimentary deposits reappear south of the Alps, and in the opinion of some high authorities, as, for instance, of Bonney and Heim, passed continuously over the intervening regions. The last great upheaval commenced after the Miocene period, and continued through the Pliocene. Miocene strata attain in the Righi a height of 6000 feet.

For neither the hills nor the mountains are everlasting. or of the same age.

The Welsh mountains are older than the Vosges, the Vosges than the Pyrenees, the Pyrenees than the Alps, and the Alps than the Andes, which indeed are still rising ; so that

if our English mountains are less imposing
so far as mere height is concerned, they are
most venerable from their great antiquity.

But though the existing Alps are in one
sense, and speaking geologically, very recent,
there is strong reason for believing that there
was a chain of lofty mountains there long
previously. "The first indication," says Judd,
"of the existence of a line of weakness in this
portion of the earth's crust is found towards
the close of the Permian period, when a series
of volcanic outbursts on the very grandest
scale took place" along a line nearly following that of the present Alps, and led to the
formation of a range of mountains, which, in
his opinion, must have been at least 8000 to
9000 feet high. Ramsay and Bonney have
also given strong reasons for believing
that the present line of the Alps was, at a
still earlier period, occupied by a range
of mountains no less lofty than those of
to-day. Thus then, though the present Alps
are comparatively speaking so recent, there
are good grounds for the belief that they were
preceded by one or more earlier ranges, once

as lofty as they are now, but which were more or less completely levelled by the action of air and water, just as is happening now to the present mountain ranges.

Movements of elevation and subsidence are still going on in various parts of the world. Scandinavia is rising in the north, and sinking at the south. South America is rising on the west and sinking in the east, rotating in fact on its axis, like some stupendous pendulum.

The crushing and folding of the strata to which mountain chains are due, and of which the Alps afford such marvellous illustrations, necessarily give rise to Earthquakes, and the slight shocks so frequent in parts of Switzerland[1] appear to indicate that the forces which have raised the Alps are not yet entirely spent, and that slow subterranean movements are still in progress along the flanks of the mountains.

But if the mountain chains are due to compression, the present valleys are mainly the result of denudation. As soon as a mountain range is once raised, all nature seems to con-

[1] In the last 150 years more than 1000 are recorded.

spire against it. Sun and Frost, Heat and Cold, Air and Water, Ice and Snow, every plant, from the Lichen to the Oak, and every animal, from the Worm to Man himself, combine to attack it. Water, however, is the most powerful agent of all. The autumn rains saturate every pore and cranny; the water as it freezes cracks and splits the hardest rocks; while the spring sun melts the snow and swells the rivers, which in their turn carry off the debris to the plains.

Perhaps, however, it would after all be more correct to say that Nature, like some great artist, carves the shapeless block into form, and endows the rude mass with life and beauty.

"What more," said Hutton long ago, " is required to explain the configuration of our mountains and valleys? Nothing but time. It is not any part of the process that will be disputed; but, after allowing all the parts, the whole will be denied; and for what? Only because we are not disposed to allow that quantity of time which the absolution of so much wasted mountain might require."

The tops of the Swiss mountains stand,

and since their elevation have probably always stood, above the range of ice, and hence their bold peaks. In Scotland, on the contrary, and still more in Norway, the sheet of ice which once, as is the case with Greenland now, spread over the whole country, has shorn off the summits and reduced them almost to gigantic bosses; while in Wales the same causes, together with the resistless action of time — for, as already mentioned, the Welsh hills are far older than the mountains of Switzerland — has ground down the once lofty summits and reduced them to mere stumps, such as, if the present forces are left to work out their results, the Swiss mountains will be thousands, or rather tens of thousands, of years hence.

The " snow line " in Switzerland is generally given as being between 8500 and 9000 feet. Above this level the snow or *névé* gradually accumulates until it forms " glaciers," solid rivers of ice which descend more or less far down the valleys. No one who has not seen a glacier can possibly realise

Fig. 20. — Glacier of the Blümlis Alp.

THE MER DE GLACE.

To face page 258.

what they are like. Fig. 20 represents the glacier of the Blümlis Alp, and the Plate the Mer de Glace.

They are often very beautiful. " Mount Beerenberg," says Lord Dufferin, " in size, colour, and effect far surpassed anything I had anticipated. The glaciers were quite an unexpected element of beauty. Imagine a mighty river, of as great a volume as the Thames, started down the side of a mountain, bursting over every impediment, whirled into a thousand eddies, tumbling and raging on from ledge to ledge in quivering cataracts of foam, then suddenly struck rigid by a power so instantaneous in its action that even the froth and fleeting wreaths of spray have stiffened to the immutability of sculpture. Unless you had seen it, it would be almost impossible to conceive the strangeness of the contrast between the actual tranquillity of these silent crystal rivers and the violent descending energy impressed upon their exterior. You must remember too all this is upon a scale of such prodigious magnitude, that when we suc-

ceeded subsequently in approaching the spot
— where with a leap like that of Niagara
one of these glaciers plunges down into the
sea — the eye, no longer able to take in its
fluvial character, was content to rest in
simple astonishment at what then appeared
a lucent precipice of grey-green ice, rising
to the height of several hundred feet above
the masts of the vessel." [1]

The cliffs above glaciers shower down
fragments of rock which gradually accumulate at the sides and at the end of
the glaciers, forming mounds known as
" moraines." Many ancient moraines occur
far beyond the present region of glaciers.

In considering the condition of alpine
valleys we must remember that the glaciers
formerly descended much further than they
do at present. The glaciers of the Rhone
for instance occupied the whole of the Valais,
filled the Lake of Geneva — or rather the
site now occupied by that lake — and rose
2000 feet up the slopes of the Jura; the
Upper Ticino, and contributory valleys, were

---

[1] *Letters from High Latitudes.*

occupied by another which filled the basin
of the Lago Maggiore ; a third occupied the
valley of the Dora Baltea, and has left a
moraine at Ivrea some twenty miles long, and
which rises no less than 1500 feet above the
present level of the river. The Scotch and
Scandinavian valleys were similarly filled
by rivers of ice, which indeed at one time
covered the whole country with an immense
sheet, as Greenland is at present. Enormous blocks of stone, the Pierre à Niton
at Geneva and the Pierre à Bot above
Neuchâtel, for instance, were carried by
these glaciers for miles and miles ; and many
of the stones in the Norfolk cliffs were
brought by ice from Norway (perhaps, however, by Icebergs), across what is now the
German Ocean. Again wherever the rocks
are hard enough to have withstood the
weather, we find them polished and ground,
just as, and even more so than, those at the
ends and sides of existing glaciers.

The most magnificent glacier tracks in the
Alps are, in Ruskin's opinion, those on the
rocks of the great angle opposite Martigny ;

the most interesting those above the channel
of the Trient between Valorsine and the valley
of the Rhone.

In Great Britain I know no better illustration of ice action than is to be seen on the
road leading down from Glen Quoich to Loch
Hourn. one of the most striking examples of
desolate and savage scenery in Scotland.   Its
name in Celtic is said to mean the Lake of
Hell.   All along the roadside are smoothed
and polished hummocks of rock, most of them
deeply furrowed with approximately parallel
striæ. presenting a gentle slope on the upper
end. and a steep side below. clearly showing
the direction of the great ice flow.

Many of the upper Swiss valleys contain
lakes, as, for instance. that of the Upper
Rhone, the Lake of Geneva. of the Reuss. the
Lake of Lucerne. of the Rhine. that of Constance.   These lakes are generally very deep.

The colour of the upper rivers. which are
white with the diluvium from the glaciers, is
itself evidence of the erosive powers which
they exercise.   This finely-divided matter is.
however, precipitated in the lakes, which, as

well as the rivers issuing from them, are a
beautiful rich blue.

"Is it not probable that this action of
finely-divided matter may have some influence on the colour of some of the Swiss lakes
—as that of Geneva for example? This lake
is simply an expansion of the river Rhone,
which rushes from the end of the Rhone
glacier, as the Arveiron does from the end of
the Mer de Glace. Numerous other streams
join the Rhone right and left during its
downward course; and these feeders, being
almost wholly derived from glaciers, join the
Rhone charged with the finer matter which
these in their motion have ground from the
rocks over which they have passed. But the
glaciers must grind the mass beneath them
to particles of all sizes, and I cannot help
thinking that the finest of them must remain
suspended in the lake throughout its entire
length. Faraday has shown that a precipi-
tate of gold may require months to sink to
the bottom of a bottle not more than five
inches high, and in all probability it would
require ages of calm subsidence to bring all

the particles which the Lake of Geneva contains to its bottom. It seems certainly worthy of examination whether such particles suspended in the water contribute to the production of that magnificent blue which has excited the admiration of all who have seen it under favourable circumstances."[1]

Among the Swiss mountains themselves each has its special character. Tyndall thus describes a view in the Alps, certainly one of the most beautiful — that, namely, from the summit of the Ægischhorn.

"Skies and summits are to-day without a cloud, and no mist or turbidity interferes with the sharpness of the outlines. Jungfrau, Monk, Eiger, Trugberg, cliffy Strahlgrat, stately lady-like Aletschhorn, all grandly pierce the empyrean. Like a Saul of Mountains, the Finsteraarhorn overtops all his neighbours; then we have the Oberaarhorn, with the riven glacier of Viesch rolling from his shoulders. Below is the Mârjelin See, with its crystal precipices and its floating icebergs, snowy white, sailing on a blue green

---

[1] *Glaciers of the Alps.*

sea. Beyond is the range which divides the
Valais from Italy. Sweeping round, the
vision meets an aggregate of peaks which look
as fledglings to their mother towards the
mighty Dom. Then come the repellent crags
of Mont Cervin; the ideal of moral savagery,
of wild untameable ferocity, mingling involun-
tarily with our contemplation of the gloomy
pile. Next comes an object, scarcely less
grand, conveying, it may be, even a deeper
impression of majesty and might than the
Matterhorn itself — the Weisshorn, perhaps
the most splendid object in the Alps. But
beauty is associated with its force, and we
think of it, not as cruel, but as grand and
strong. Further to the right the great
Combin lifts up his bare head; other peaks
crowd around him; while at the extremity of
the curve round which our gaze has swept
rises the sovran crown of Mont Blanc. And
now, as day sinks, scrolls of pearly clouds
draw themselves around the mountain crests,
being wafted from them into the distant air.
They are without colour of any kind; still, by
grace of form, and as the embodiment of

lustrous light and most tender shade, their beauty is not to be described."[1]

VOLCANOES

Volcanoes belong to a totally different series of mountains.

It is practically impossible to number the Volcanoes on our earth. Humboldt enumerated 223, which Keith Johnston raised to nearly 300. Some, no doubt, are always active, but in the majority the eruptions are occasional, and though some are undoubtedly now extinct, it is impossible in all cases to distinguish those which are only in repose from those whose day of activity is over. Then, again, the question would arise, which should be regarded as mere subsidiary cones and which are separate volcanoes. The slopes of Etna present more than 700 small cones, and on Hawaii there are several thousands. In fact, most of the very lofty volcanoes present more or less lateral cones.

The molten matter, welling up through

[1] *Mountaineering in* 1861.

some fissure, gradually builds itself up into a cone, often of the most beautiful regularity, such as the gigantic peaks of Chimporazo, Cotopaxi (Fig. 21), and Fusiyama, and hence it is that the crater is so often at, or very near, the summit.

Perhaps no spectacle in Nature is more magnificent than a Volcano in activity. It has been my good fortune to have stood

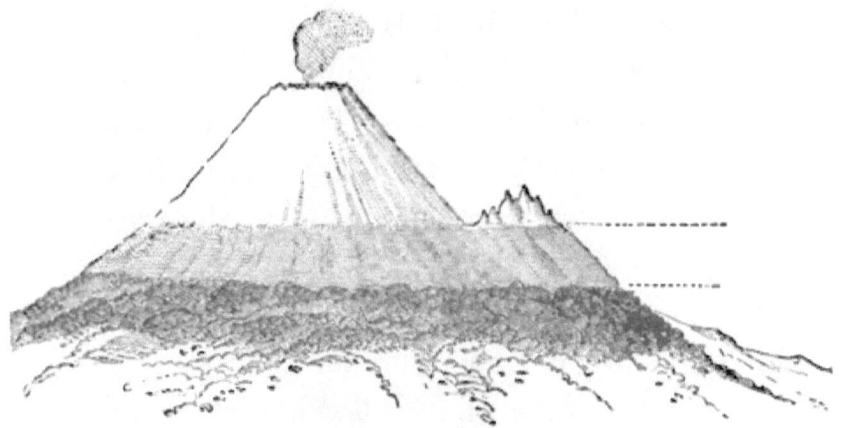

Fig. 21.—Cotopaxi.

more than once at the edge of the crater of Vesuvius during an eruption, to have watched the lava seething below, while enormous stones were shot up high into the air. Such a spectacle can never be forgotten.

The most imposing crater in the world is probably that of Kilauea, at a height of about 4000 feet on the side of Mouna Loa, in the Island of Hawaii. It has a diameter of 2 miles, and is elliptic in outline, with a longer axis of about 3, and a circumference of about 7 miles. The interior is a great lake of lava, the level of which is constantly changing. Generally, it stands about 800 feet below the edge, and the depth is about 1400 feet. The heat is intense, and, especially at night, when the clouds are coloured scarlet by the reflection from the molten lava, the effect is said to be magnificent. Gradually the lava mounts in the crater until it either bursts through the side or runs over the edge, after which the crater remains empty, sometimes for years.

A lava stream flows down the slope of the mountain like a burning river, at first rapidly, but as it cools, scoriæ gradually form, and at length the molten matter covers itself completely (Fig. 22), both above and at the sides, with a solid crust, within which, as in a tunnel, it continues to flow

Fig. 22. — Lava Stream.

slowly as long as it is supplied from the
source, here and there breaking through the
crust which, as continually, re-forms in front.
Thus the terrible, inexorable river of fire
slowly descends, destroying everything in
its course.

The stream of lava which burst from
Mouna Loa in 1885 had a length of 70 miles;
that of Skaptar-Jokul in Iceland in 1783 had
a length of 50 miles, and a maximum depth
of nearly 500 feet. It has been calculated that
the mass of lava equalled that of Mont Blanc.

The stones, ashes, and mud ejected during
eruptions are even more destructive than the
rivers of lava. In 1851 Tomboro, a volcano
on the Island of Sumbava, cost more lives
than fell in the battle of Waterloo. The
earthquake of Lisbon in 1755 destroyed
60,000 persons. During the earthquake of
Riobamba and the mud eruption of Tungu-
ragua, and again in that of Krakatoa, it is
estimated that the number who perished was
between 30,000 and 40,000. At the earthquake of Antioch in 526 no less than 200,000
persons are said to have lost their lives.

Perhaps the most destructive eruption of modern times has been that on Cosequina. For 25 miles it covered the ground with muddy water 16 feet in depth. The dust and ashes formed a dense cloud, extending over many miles, some of it being carried 20 degrees to the west. The total mass ejected has been estimated at 60 milliards of square yards.

Stromboli, in the Mediterranean (Fig. 23), though only 2500 feet in height, is very imposing from its superb regularity, and its roots plunge below the surface to a depth of 4000 feet.

It is, moreover, very interesting from the regularity of its action, which has a period of 5 minutes or a little less. On looking down into the crater one sees at a depth of say 300 feet a seething mass of red-hot lava; this gradually rises, and then explodes, throwing up a cloud of vapour and stones, after which it sinks again. So regular is it that the Volcano has been compared to a "flashing" lighthouse, and this wonderful process has been going on for ages.

R

Fig. 23. — Stromboli, viewed from the north-west, April 1874.

Though long extinct, volcanoes once existed in the British Isles; Arthur's Seat, near Edinburgh, for instance, appears to be the funnel of a small volcano, belonging to the Carboniferous period.

The summit of a volcanic mountain is sometimes entirely blown away. Between my first two visits to Vesuvius 200 feet of the mountain had thus disappeared. Vesuvius itself stands in a more ancient crater, part of which still remains, and is now known as Somma, the greater portion having disappeared in the great eruption of 79, when the mountain, waking from its long sleep, destroyed Herculaneum and Pompeii.

As regards the origin of volcanoes there have been two main theories. Impressed by the magnitude and grandeur of the phenomena, enhanced as they are by their destructive character, many have been disposed to regard the craters of volcanoes as gigantic chimneys, passing right through the solid crust of the globe, and communicating with a central fire. Recent researches, however, have indicated that, grand and imposing as they are, volcanoes must yet be regarded as due mainly to local and superficial causes.

A glance at the map shows that volcanoes are almost always situated on, or near, the sea coast. From the interior of continents they are entirely wanting. The number of active volcanoes in the Andes, contrasted with their absence in the Alps and Ourals, the Himalayas, and Central Asian chains, is very striking. Indeed, the Pacific Ocean is encircled, as Ritter has pointed out, by a ring of fire. Beginning with New Zealand, we have the Volcanoes of Tongariro, Whakaii, etc.; thence the circle passes through the Fiji Islands. Solomon Islands, New Guinea, Timor, Flores, Sumbava, Lombock, Java, Sumatra, the Philippines, Japan, the Aleutian Islands, along the Rocky Mountains, Mexico, Peru, and Chili, to Tierra del Fuego, and, in the far south, to the two great Volcanoes of Erebus and Terror on Victoria Land.

We know that the contraction of the Earth's surface with the strains and fractures, the compression and folds, which must inevitably result, is still in operation, and must give rise to areas of high temperature, and consequently to volcanoes. We must also remember that the real mountain chains of our earth are the continents, compared to which even the Alps and Andes are mere wrinkles. It is along the lines of the great mountain chains, that is to say, along the main coast lines, rather than in the centres of the continents, which may be regarded as comparatively quiescent, that we should naturally expect to find the districts of greatest heat, and this is perhaps why volcanoes are generally distributed along the coast lines.

Another reason for regarding Volcanoes as local phenomena is that many even of those comparatively near one another act quite independently. This is so with Kilauea and Mouna Loa, both on the small island of Hawaii.

Again, if volcanoes were in connection with a great central sea of fire, the eruptions must follow the same laws as regulate the tides. This, however, is not the case. There are indeed indications of the existence of slight tides in the molten lake which underlies Vesuvius, and during the eruption of 1865 there was increased activity twice a day, as we should expect to find in any great fluid reservoir, but very different indeed from what must have been the case if the mountain was in connection with a central ocean of molten matter.

Indeed, unless the "crust" of our earth was of great thickness we should be subject to perpetual earthquakes. No doubt these are far more frequent than is generally supposed; indeed, with our improved instruments it can be shown that instead of occasional vibrations, with long intermediate periods of rest, we have in reality short intervals of rest with long periods of vibration, or rather perhaps that the crust of the earth is in constant tremor, with more violent oscillation from time to time.

It appears, moreover, that earthquakes are not generally deep-seated. The point at which the shock is vertical can be ascertained, and it is also possible in some cases to determine the angle at which it emerges elsewhere. When this has been done it has always been found that the seat of disturbance must have been within 30 geographical miles of the surface.

Yet, though we cannot connect volcanic action with the central heat of the earth, but must regard it as a minor and local manifestation of force, volcanoes still remain among the grandest, most awful, and at the same time most magnificent spectacles which the earth can afford.

CHAPTER VII

WATER

Of all inorganic substances, acting in their own proper nature, and without assistance or combination, water is the most wonderful. If we think of it as the source of all the changefulness and beauty which we have seen in the clouds; then as the instrument by which the earth we have contemplated was modelled into symmetry, and its crags chiselled into grace; then as, in the form of snow, it robes the mountains it has made, with that transcendent light which we could not have conceived if we had not seen; then as it exists in the foam of the torrent, in the iris which spans it, in the morning mist which rises from it, in the deep crystalline pools which mirror its hanging shore, in the broad lake and glancing river, finally, in that which is to all human minds the best emblem of unwearied, unconquerable power, the wild, various, fantastic, tameless unity of the sea; what shall we compare to this mighty, this universal element, for glory and for beauty? or how shall we follow its eternal cheerfulness of feeling? It is like trying to paint a soul. — Ruskin.

RYDAL WATER.

To face page 251.

CHAPTER VII

WATER

In the legends of ancient times running water was proof against all sorcery and witchcraft :

> No spell could stay the living tide
> Or charm the rushing stream.[1]

There was much truth as well as beauty in this idea.

Flowing waters, moreover, have not only power to wash out material stains, but they also clear away the cobwebs of the brain — the results of over incessant work — and restore us to health and strength.

Snowfields and glaciers, mountain torrents, sparkling brooks, and stately rivers, meres and lakes, and last, not least, the great ocean itself, all alike possess this magic power.

[1] Leyden.

"When I would beget content," says Izaak
Walton, "and increase confidence in the
power and wisdom and providence of Al-
mighty God, I will walk the meadows by
some gliding stream, and there contemplate
the lilies that take no care, and those very
many other little living creatures that are
not only created, but fed (man knows not
how) by the goodness of the God of Nature,
and therefore trust in Him;" and in his
quaint old language he craves a special blessing on all those "that are true lovers of
virtue, and dare trust in His Providence, and
be quiet, and go a angling."

At the water's edge flowers are especially
varied and luxuriant, so that the banks of a
river are a long natural garden of tall and
graceful grasses and sedges, the Meadow
Sweet, the Flowering Rush, the sweet Flag,
the Bull Rush, Purple Loosestrife, Hemp
Agrimony, Dewberry, Forget-me-not, and a
hundred more, backed by Willows, Alders,
Poplars, and other trees.

The Animal world, if less conspicuous to
the eye, is quite as fascinating to the imagina-

tion. Here and there a speckled Trout may
be detected (rather by the shadow than the
substance) suspended in the clear water, or
darting across a shallow; if we are quiet we
may see Water Hens or Wild Ducks swimming among the lilies, a Kingfisher sitting on
a branch or flashing away like a gleam of
light; a solemn Heron stands maybe at the
water's edge, or slowly rises flapping his
great wings; Water Rats, neat and clean
little creatures, very different from their
coarse brown namesakes of the land, are
abundant everywhere; nor need we even yet
quite despair of seeing the Otter himself.

Insects of course are gay, lively, and innumerable; but after all the richest fauna is
that visible only with a microscope.

"To gaze," says Dr. Hudson, "into that
wonderful world which lies in a drop of
water, crossed by some stems of green weed,
to see transparent living mechanism at work,
and to gain some idea of its modes of action,
to watch a tiny speck that can sail through
the prick of a needle's point; to see its
crystal armour flashing with ever varying

tint, its head glorious with the halo of its quivering cilia ; to see it gliding through the emerald stems, hunting for its food, snatching at its prey, fleeing from its enemy, chasing its mate (the fiercest of our passions blazing in an invisible speck); to see it whirling in a mad dance, to the sound of its own music, the music of its happiness, the exquisite happiness of living — can any one, who has once enjoyed this sight, ever turn from it to mere books and drawings, without the sense that he has left all Fairyland behind him?" [1]

The study of Natural History has indeed the special advantage of carrying us into the country and the open air.

Lakes are even more restful than rivers or the sea. Rivers are always flowing, though it may be but slowly; the sea may rest awhile, now and then, but is generally full of action and energy; while lakes seem to sleep and dream. Lakes in a beautiful country are like silver ornaments on a lovely dress, like liquid gems in a beautiful setting, or bright eyes in a lovely face. Indeed as we gaze

[1] Dr. Hudson, Address to the Microscopical Society, 1889.

WINDERMERE.

To face page 254.

down on a lake from some hill or cliff it almost looks solid, like some great blue crystal.

It is not merely for purposes of commerce or convenience that men love to live near rivers.

> Let me live harmlessly, and near the brink
>   Of Trent or Avon have my dwelling-place;
> Where I may see my quill, or cork, down sink,
>   With eager bite of pike, or bleak, or dace;
> And on the world and my Creator think:
>   While some men strive ill-gotten goods t' embrace:
> And others spend their time in base excess
> Of wine; or worse, in war, or wantonness.
>
> Let them that will, these pastimes still pursue,
>   And on such pleasing fancies feed their fill:
> So I the fields and meadows green may view
>   And daily by fresh rivers walk at will,
> Among the daisies and the violets blue,
>   Red hyacinth and yellow daffodil.[1]

It is interesting and delightful to trace a river from its source to the sea.

"Beginning at the hill-tops," says Geikie, "we first meet with the spring or 'well-eye,' from which the river takes its rise. A patch of bright green, mottling the brown heathy

[1] F. Davors.

slope, shows where the water comes to the
surface, a treacherous covering of verdure
often concealing a deep pool beneath. From
this source the rivulet trickles along the grass
and heath, which it soon cuts through, reaching the black, peaty layer below, and running
in it for a short way as in a gutter. Excavating its channel in the peat, it comes down
to the soil, often a stony earth bleached white
by the peat. Deepening and widening the
channel as it gathers force with the increasing slope, the water digs into the coating of
drift or loose decomposed rock that covers
the hillside. In favourable localities a narrow precipitous gully, twenty or thirty feet
deep, may thus be scooped out in the course
of a few years."

If, however, we trace one of the Swiss
rivers to its source we shall generally find
that it begins in a snow field or _névé_ nestled
in a shoulder of some great mountain.

Below the _névé_ lies a glacier, on, in, and
under which the water runs in a thousand
little streams, eventually emerging at the
end, in some cases forming a beautiful blue

cavern, though in others the end of the glacier is encumbered and concealed by earth and stones.

Fig. 24. — Upper Valley of St. Gotthard.

The uppermost Alpine valleys are perhaps generally, though by no means always, a little desolate and severe, as, for instance, that of St. Gotthard (Fig. 24). The sides are clothed with rough pasture, which is flowery indeed, though of course the flowers are not visible at a distance, interspersed with live rock and fallen masses, while along the bottom rushes a white torrent. The snowy peaks are generally more or less hidden by the shoulders of the hills.

The valleys further down widen and become more varied and picturesque. The snowy peaks and slopes are more often visible, the " alps " or pastures to which the cows are taken in summer, are greener and dotted with the huts or châlets of the cowherds, while the tinkling of the cowbells comes to one from time to time, softened by distance, and suggestive of mountain rambles. Below the alps there is generally a steeper part clothed with Firs or with Larches and Pines, some of which seem as if they were scaling the mountains in regiments, preceded by a certain number of skirmishers. Below the fir woods again are Beeches, Chestnuts, and other deciduous trees, while the central

cultivated portion of the valley is partly
arable, partly pasture, the latter differing
from our meadows in containing a greater
variety of flowers — Campanulas, Wild Geraniums, Chervil, Ragged Robin, Narcissus,
etc. Here and there is a brown village,
while more or less in the centre hurries
along, with a delightful rushing sound, the
mountain torrent, to which the depth, if not
the very existence of the valley, is mainly
due. The meadows are often carefully
irrigated, and the water power is also used
for mills, the streams seeming to rush on, as
Ruskin says, "eager for their work at the
mill, or their ministry to the meadows."

Apart from the action of running water,
snow and frost are continually disintegrating
the rocks, and at the base of almost any
steep cliff may be seen a slope of debris
(as in Figs. 25, 26). This stands at a regular
angle — the angle of repose — and unless it
is continually removed by a stream at the
base, gradually creeps up higher and higher,
until at last the cliff entirely disappears.

Sometimes the two sides of the valley

approach so near that there is not even room
for the river and the road: in that case
Nature claims the supremacy, and the road
has to be carried in a cutting, or perhaps in
a tunnel through the rock. In other cases
Nature is not at one with herself. In many

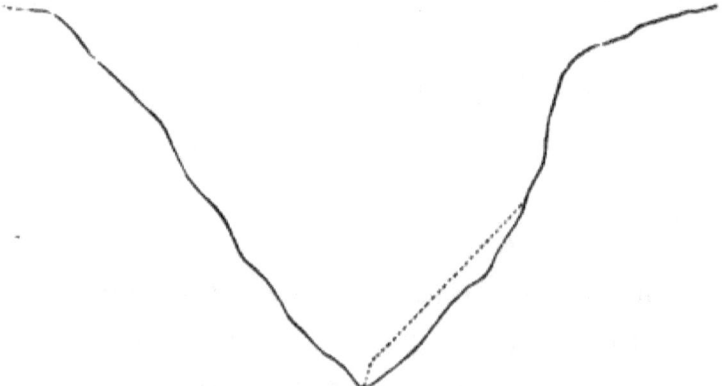

Fig. 25. — Section of a river valley. The dotted line shows a slope or
talus of debris.

places the debris from the rocks above would
reach right across the valley and dam up the
stream. Then arises a struggle between
rock and river, but the river is always victorious in the end; even if dammed back for
a while, it concentrates its forces, rises up
the rampart of rock, rushes over trium-
phantly, resumes its original course, and
gradually carries the enemy away.

Another prominent feature in many valleys is afforded by the old river, or lake, terraces,

Fig. 26. — Valley of the Rhone, with the waterfall of Sallenches, showing talus of debris.

which were formed at a time when the river ran at a level far above its present bed.

Thus many a mountain valley gives some such section as the following.

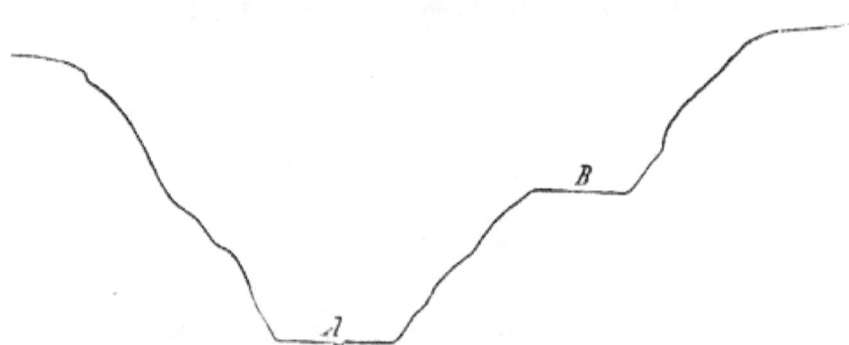

Fig. 27. — *A*, present river valley; *B*, old river terrace.

First, a face of rock, very steep, and in some places almost perpendicular; secondly, a regular talus of fallen rocks, stones, etc., as shown in the view of the Rhone Valley (Fig. 26), which takes what is known as the slope of repose, at an angle which depends on the character of the material. As a rule for loose rock fragments it may be taken roughly to be an angle of about 45°. Then an irregular slope followed in many places by one or more terraces, and lastly the present bed of the river.

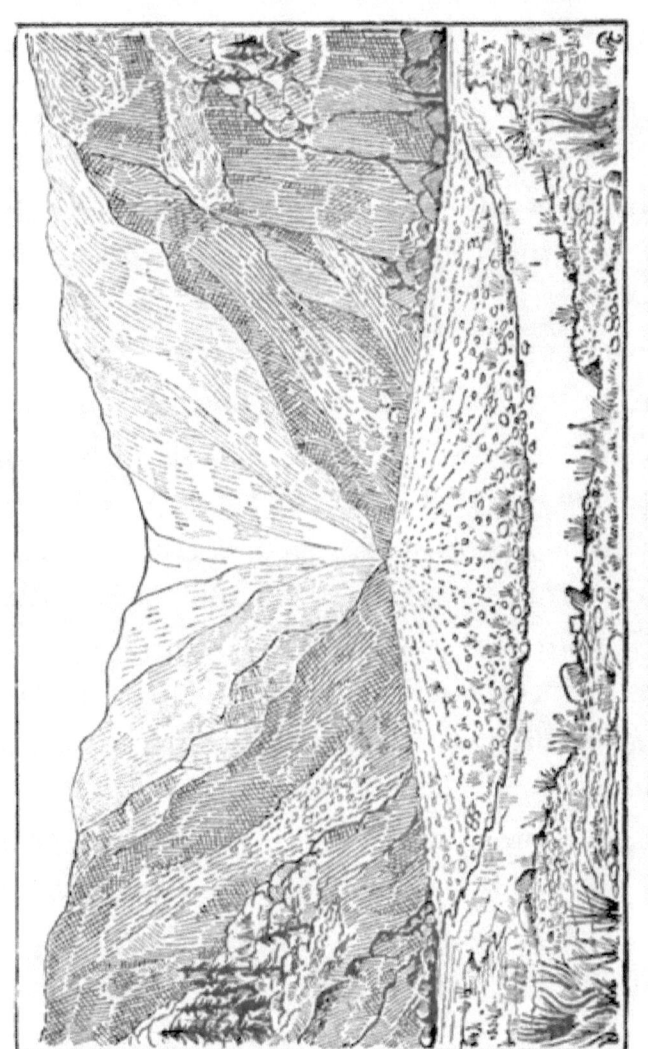

Fig. 28. — Diagram of an Alpine valley showing a river cone.   Front view.

The width or narrowness of the valley in relation to its depth depends greatly on the condition of the rocks, the harder and tougher they are the narrower as a rule being the valley.

From time to time a side stream enters the main valley. This is itself composed of many smaller rivulets. If the lateral valleys are steep, the streams bring with them, especially after rains, large quantities of earth and stones. When, however, they reach the main valley, the rapidity of the current being less, their power of transport also diminishes, and they spread out the material which they carry down in a depressed cone (Figs. 28, 29, 31, 32).

A side stream with its terminal cone, when seen from the opposite side of the valley, presents the appearance shown in Figs. 28, 31, or, if we are looking down the valley, as in Figs. 29, 32, the river being often driven to one side of the main valley, as, for instance, is the case in the Valais, near Sion, where the Rhone (Fig. 30) is driven out of its course by, and forms a curve round, the cone brought down by the torrent of the Borgne.

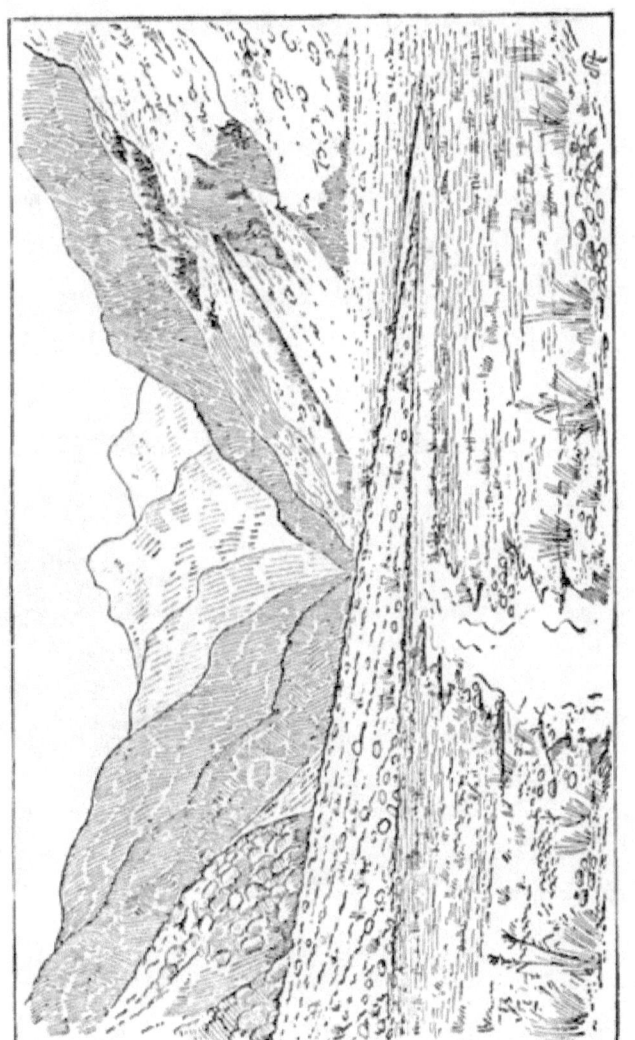

Fig. 29. — Diagram of an Alpine valley, showing a river cone. Lateral view.

Sometimes two lateral valleys (see Plate) come down nearly opposite one another, so that the cones meet, as, for instance, some little way below Vernayaz, and, indeed, in several other places in the Valais (Fig. 31). Or more permanent lakes may be due to a

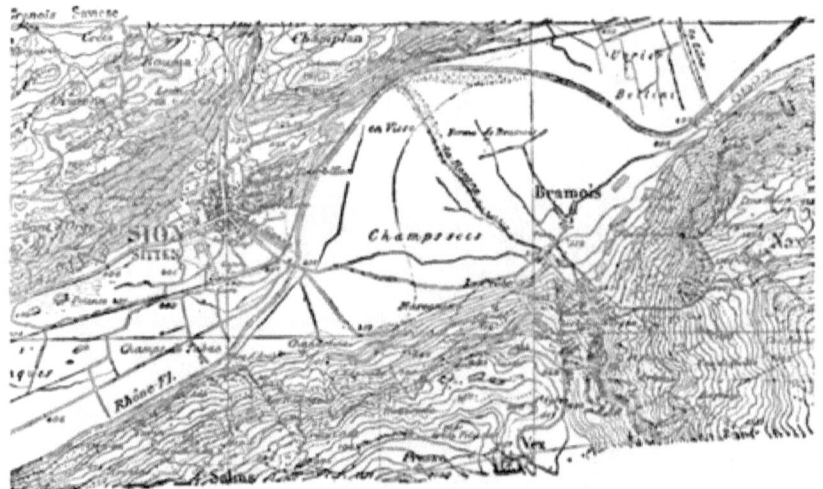

Fig. 30.

ridge of rock running across the valley, as, for instance, just below St. Maurice in the Valais.

Almost all river valleys contain, or have contained, in their course one or more lakes, and where a river falls into a lake a cone like

VIEW IN THE VALAIS BELOW ST. MAURICE.

To face page 266.

those just described is formed, and projects
into the lake.   Thus on the Lake of Geneva,

Fig. 31. — View in the Rhone Valley, showing a lateral cone.

between Vevey and Villeneuve (see Fig. 33).
there are several such promontories, each

marking the place where a stream falls into the lake.

Fig. 32. — View in the Rhone Valley, showing the slope of a river cone.

The Rhone itself has not only filled up what was once the upper end of the lake,

but has built out a strip of land into the
water.

Fig. 33.—Shore of the Lake of Geneva, near Vevey.

That the lake formerly extended some
distance up the Valais no one can doubt
who looks at the flat ground about Ville-

neuve. The Plate opposite, from a photograph taken above Vevey, shows this clearly. It is quite evident that the lake must formerly have extended further up the valley, and that it has been filled up by material brought down by the Rhone, a process which is still continuing.

At the other end of the lake the river rushes out 15 feet deep of "not flowing, but flying water; not water neither — melted glacier matter, one should call it; the force of the ice is in it, and the wreathing of the clouds, the gladness of the sky, and the countenance of time." [1]

In flat countries the habits of rivers are very different. For instance, in parts of Norfolk there are many small lakes or "broads" in a network of rivers — the Bure, the Yare, the Ant, the Waveney, etc. — which do not rush on with the haste of some rivers, or the stately flow of others which are steadily set to reach the sea, but rather seem like rivers wandering in the meadows on a holiday. They have often no natural banks, but are

[1] Ruskin.

VIEW UP THE VALAIS FROM THE LAKE OF GENEVA.

To face page 270.

bounded by dense growths of tall grasses,
Bulrushes, Reeds, and Sedges, interspersed
with the spires of the purple Loosestrife,
Willow Herb, Hemp Agrimony, and other

Fig. 34. — View in the district of the Broads, Norfolk.

flowers, while the fields are very low and protected by dykes, so that the red cattle appear
to be browsing below the level of the water;
and as the rivers take most unexpected
turns, the sailing boats often seem (Fig. 34)
as if they were in the middle of the fields.

At present these rivers are restrained in
their courses by banks; when left free they

are continually changing their beds. Their
courses at first sight seem to follow no rule,
but, as it is termed, from a celebrated river
of Asia Minor, to " meander " along without
aim or object, though in fact they follow
very definite laws.

Finally, when the river at length reaches
the sea, it in many cases spreads out in the
form of a fan, forming a very flat cone or
" delta," as it is called, from the Greek capi-
tal Δ, a name first applied to that of the
Nile, and afterwards extended to other rivers.
This is due to the same cause, and resembles,
except in size, the comparatively minute
cones of mountain streams.

Fig. 35 represents the delta of the Po, and
it will be observed that Adria, once a great
port, and from which the Adriatic was named,
is now more than 20 miles from the sea.
Perhaps the most remarkable case is that of
the Mississippi (Fig. 36), the mouths of which
project into the sea like a hand, or like the
petals of a flower.. For miles the mud is too
soft to support trees, but is covered by sedges
(Miegea) ; the banks of mud gradually be-

come too soft and mobile even for them.
The pilots who navigate ships up the river

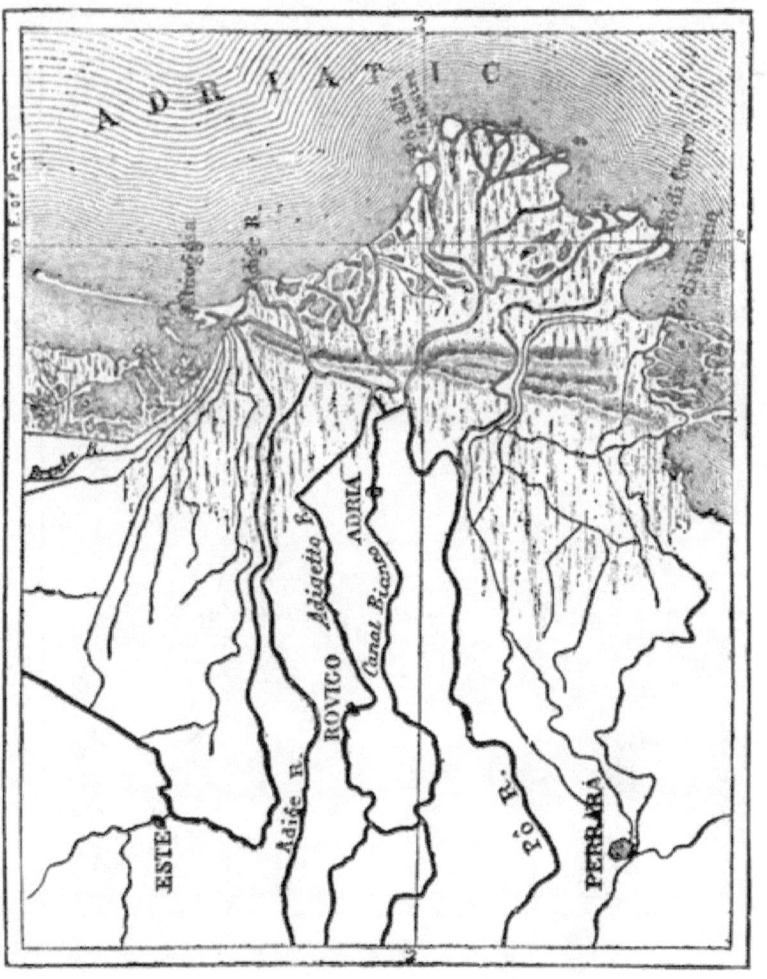

Fig. 35.

live in frail houses resting on planks, and
kept in place by anchors. Still further, and

T

the banks of the Mississippi, if banks they
can be called, are mere strips of reddish mud,

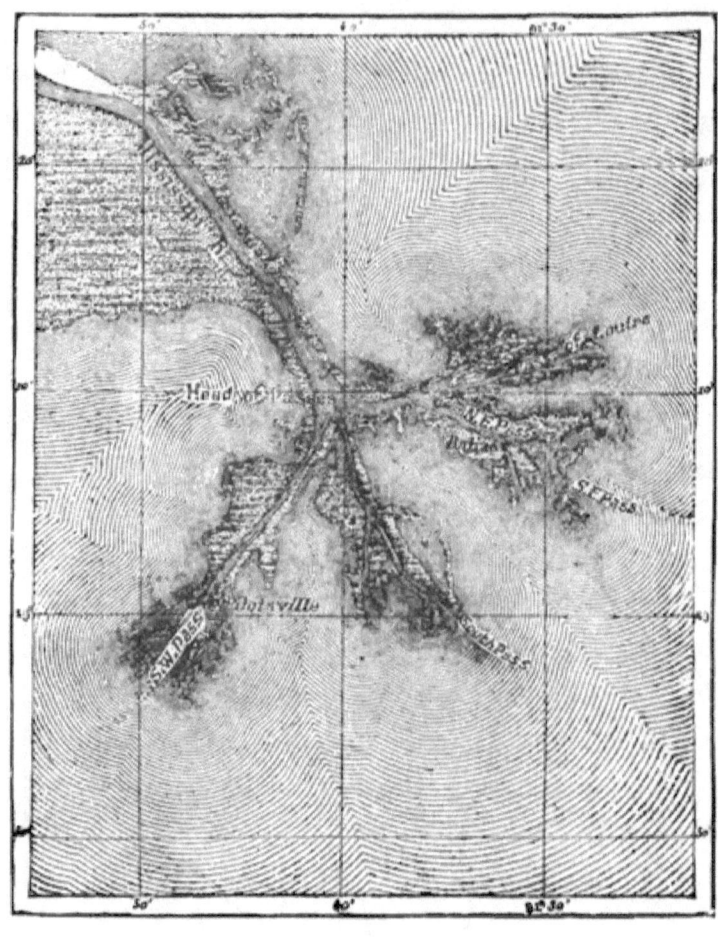

Fig. 36.

intersected from time to time by transverse
streams of water, which gradually separate

them into patches. These become more and more liquid, until the land, river, and sea merge imperceptibly into one another. The river is so muddy that it might almost be called land, and the mud so saturated by water that it might well be called sea, so that one can hardly say whether a given spot is on the continent, in the river, or on the open ocean.

CHAPTER VIII

RIVERS AND LAKES

CHAPTER VIII

RIVERS AND LAKES

ON THE DIRECTIONS OF RIVERS

IN the last chapter I have alluded to the wanderings of rivers within the limits of their own valleys; we have now to consider the causes which have determined the directions of the valleys themselves.

If a tract of country were raised up in the form of a boss or dome, the rain which fell on it would partly sink in, partly run away to the lower ground. The least inequality in the surface would determine the first directions of the streams, which would carry down any loose material, and thus form little channels, which would be gradually deepened and enlarged. It is as difficult for a river as for a man to get out of a groove.

In such a case the rivers would tend to radiate with more or less regularity from the centre or axis of the dome, as, for instance, in our English lake district (Fig. 37). Derwent Water, Thirlmere, Coniston Water, and Windermere, run approximately N. and S.; Crummock Water, Loweswater, and Buttermere N.W. by S.E.; Waste Water, Ullswater, and Hawes Water N.E. by S.W.; while Ennerdale Water lies nearly E. by W. Can we account in any way, and if so how, for these varied directions?

The mountains of Cumberland and Westmoreland form a more or less oval boss, the axis of which, though not straight, runs practically from E.N.E. to W.S.W., say from Scaw Fell to Shap Fell; and a sketch map shows us almost at a glance that Derwent Water, Thirlmere, Ullswater, Coniston Water, and Windermere run at right angles to this axis; Ennerdale Water is just where the boss ends and the mountains disappear; while Crummock Water and Waste Water lie at the intermediate angles.

So much then for the direction. We have

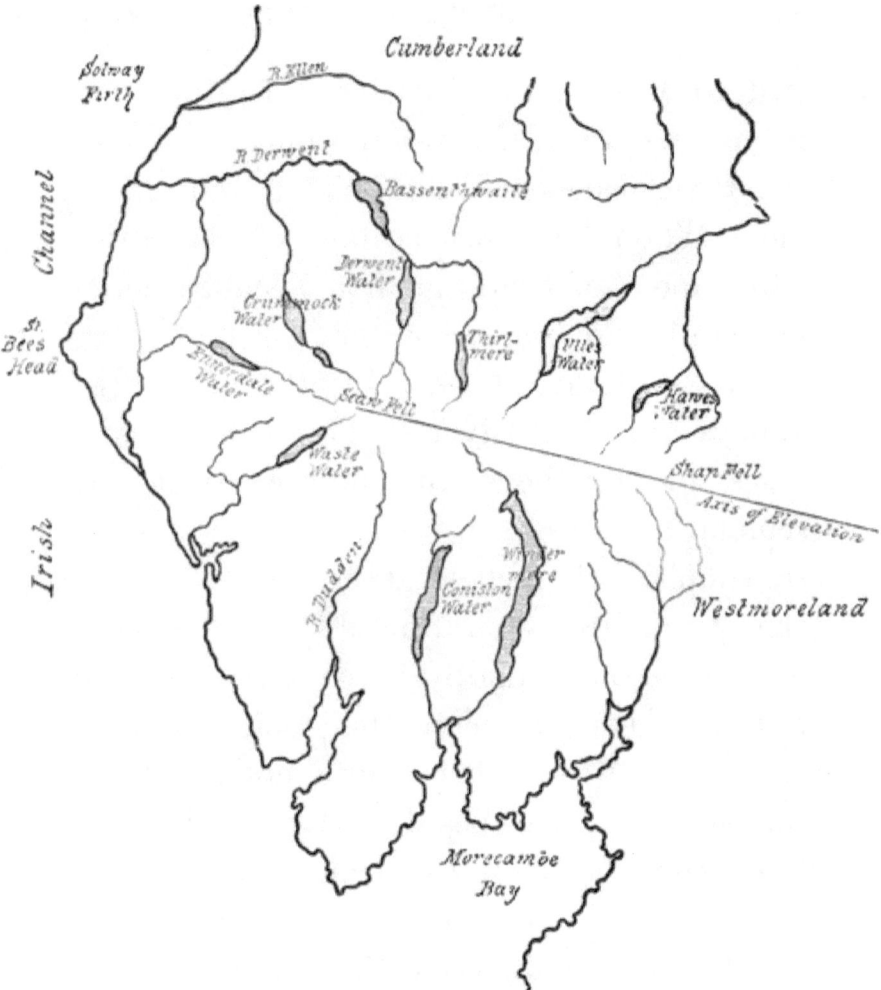

Fig. 37. — Map of the Lake District.

still to consider the situation and origin, and
it appears that Ullswater, Coniston Water,
the River Dudden, Waste Water, and Crummock Water lie along the lines of old faults,
which no doubt in the first instance determined the flow of the water.

Take another case. In the Jura the
valleys are obviously (see Fig. 18) in many
cases due to the folding of the strata. It
seldom happens, however, that the case is
so simple. If the elevation is considerable
the strata are often fractured, and fissures
are produced. Again if the part elevated
contains layers of more than one character,
this at once establishes differences. Take,
for instance, the Weald of Kent (Figs. 38,
39). Here we have (omitting minor layers)
four principal strata concerned, namely, the
Chalk, Greensand, Weald Clay, and Hastings
Sands.

The axis of elevation runs (Fig. 39) from
Winchester by Petersfield, Horsham, and
Winchelsea to Boulogne, and as shown in
the following section, taken from Professor
Ramsay, we have on each side of the axis

two ridges or "escarpments," one that of
the Chalk, the other that of the Greensand,
while between the Chalk and the Greensand is a valley, and between the Greensand and the ridge of Hastings Sand an
undulating plain, in each case with a gentle slope from about where the London and

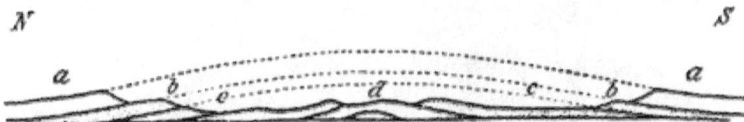

Fig. 38. — *a, a*, Upper Cretaceous strata, chiefly Chalk, forming the North
and South Downs; *b, b*, Escarpment of Lower Greensand, with a valley between it and the Chalk; *c, c*, Weald Clay, forming plains; *d*, Hills formed
of Hastings Sand and Clay. The Chalk, etc., once spread across the country,
as shown in the dotted lines.

Brighton railway crosses the Weald towards
the east. Under these circumstances we
might have expected that the streams drain-
ing the Weald would have run in the direction of the axis of elevation. and at the
bases of the escarpments, as in fact the
Rother does for part of its course, into the
sea between the North and South Downs,
instead of which as a rule they run north
and south. cutting in some cases directly
through the escarpments; on the north, for

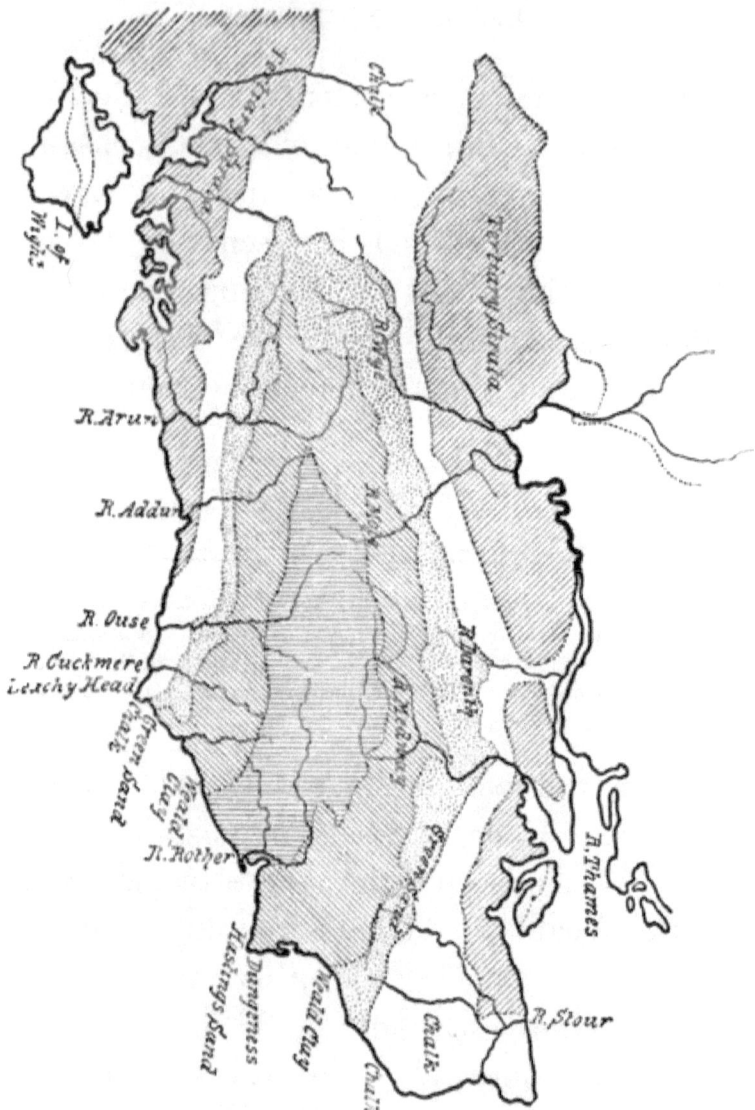

Fig. 39. — Map of the Weald of Kent.

instance, the Wye, the Mole, the Darenth, the Medway, and the Stour; and on the south the Arun, the Addur, the Ouse, and the Cuckmere.

They do not run in faults or cracks, and it is clear that they could not have excavated their present valleys under circumstances such as now exist. They carry us back indeed to a time when the Greensand and Chalk were continued across the Weald in a great dome, as shown by the dotted lines in Fig. 38. They then ran down the slope of the dome, and as the Chalk and Greensand gradually weathered back, a process still in operation, the rivers deepened and deepened their valleys, and thus were enabled to keep their original course.

Other evidence in support of this view is afforded by the presence of gravel beds in some places at the very top of the Chalk escarpment — beds which were doubtless deposited when, what is now the summit of a hill, was part of a continuous slope.

The course of the Thames offers us a somewhat similar instance. It rises on the Oolites near Cirencester, and cuts through the escarpment of the Chalk between Wallingford and Reading. The cutting through the Chalk has evidently been effected by the river itself. But this could not have happened under existing conditions. We must remember, however, that the Chalk escarpment is gradually moving eastwards. The Chalk escarpments indeed are everywhere, though of course only slowly, crumbling away. Between Farnham and Guildford the Chalk is reduced to a narrow ridge known as the Hog's Back. In the same way no doubt the area of the Chalk formerly extended much further west than it does at present. and. indeed, there can be little doubt, somewhat further west than the source of the Thames. almost to the valley of the Severn. At that time the Thames took its origin in a Chalk spring. Gradually. however. the Chalk was worn away by the action of weather. and especially of rain. The river maintained its course while gradually excavating. and sinking deeper and deeper into, the Chalk. At present the river meets the Chalk escarpment near Wallingford, but the escarpment itself is still gradually retreating eastward.

So, again, the Elbe cuts right across the Erz-Gebirge, the Rhine through the mountains between Bingen and Coblenz, the Potomac, the Susquehannah, and the Delaware through the Alleghanies. The case of the Dranse will be alluded to further on (p. 292). In these cases the rivers preceded the mountains. Indeed as soon as the land rose above the waters, rivers would begin their work, and having done so, unless the rate of elevation of the mountain exceeded the power of erosion of the river, the two would proceed simultaneously, so that the river would not alter its course, but would cut deeper and deeper as the mountain range gradually rose.

Rivers then are in many cases older than mountains. Moreover, the mountains are passive, the rivers active. Since it seems to be well established that in Switzerland a mass, more than equal to what remains, has been removed; and that many of the present mountains are not sites which were originally raised highest, but those which have suffered least, it follows that if in some cases the course of the river is due to the direction of the mountain ridges, on the other hand the direction of some of the present ridges is due to that of the rivers. At any rate it is certain that of the original surface not a trace or a fragment remains *in situ*. Many of our own English mountains were once valleys, and many of our present valleys occupy the sites of former mountain ridges.

Heim and Rütimeyer point out that of the two factors which have produced the relief of mountain regions, the one, elevation, is temporary and transitory ; the other, denudation, is constant, and gains therefore finally the upper hand.

We must not, however, expect too great regularity. The degree of hardness, the texture, and the composition of the rocks cause great differences.

On the other hand, if the alteration of level was too rapid, the result might be greatly to alter the river courses. Mr. Darwin mentions such a case, which, moreover, is perhaps the more interesting as being evidently very recent.

"Mr. Gill," he says, "mentioned to me a most interesting, and as far as I am aware, quite unparalleled case, of a subterranean disturbance having changed the drainage of a country. Travelling from Casma to Huaraz (not very far distant from Lima) he found a plain covered with ruins and marks of ancient cultivation, but now quite barren. Near it was the dry course of a considerable river, whence the water for irrigation had formerly been conducted. There was nothing in the appearance of the water-course to indicate that the river had not flowed there a few years previously; in some parts beds of sand and gravel were spread out; in others, the solid rock had been worn into a broad channel, which in one spot was about 40 yards in breadth and 8 feet deep. It is self-evident that a person following up the course of a stream will always ascend at a greater or less inclination. Mr. Gill therefore, was much astonished when walking up the bed of this ancient river, to find himself suddenly going downhill. He imagined that the downward slope had a fall of about 40 or 50 feet perpendicular. We here have unequivocal evidence that a ridge had been uplifted right across the old bed of a stream. From the moment the river course was thus arched, the water must necessarily have been thrown back, and a new channel formed. From that moment also the neighbouring plain must have lost its fertilising stream, and become a desert." [1]

The strata, moreover, often — indeed generally, as we have seen, for instance, in the case of Switzerland — bear evidence of most violent contortions, and even where the convulsions were less extreme, the valleys thus resulting are sometimes complicated by the existence of older valleys formed under previous conditions.

In the Alps then the present configuration of the surface is mainly the result of denudation. If we look at a map of Switzerland we can trace but little relation between the river courses and the mountain chains.

[1] Darwin's *Voyage of a Naturalist.*

The rivers, as a rule (Fig. 40), run either

Fig. 40. — Sketch Map of the Swiss Rivers.

S.E. by N.W., or, at right angles to this, N.E. and S.W.   The Alps themselves follow a

somewhat curved line from the Maritime Alps.
commencing with the islands of Hyères, by
Briancon, Martigny, the Valais, Urseren Thal,
Vorder Rhein, Innsbruck, Radstadt, and
Rottenmann to the Danube, a little below
Vienna.—at first nearly north and south, but
gradually curving round until it becomes
S.W. by N.E.

The central mountains are mainly composed
of Gneiss, Granite, and crystalline Schists:
the line of junction between these rocks and
the secondary and tertiary strata on the north,
runs, speaking roughly, from Hyères to Grenoble, and then by Albertville, Sion, Chur, Inns,
bruck, Radstadt, and Hieflau, towards Vienna.
It is followed (in some part of their course)
by the Isère, the Rhone, the Rhine, the Inn,
and the Enns.   One of the great folds shortly
described in the preceding chapter runs up
the Isère, along the Chamouni Valley, up the
Rhone, through the Urseren Thal, down the
Rhine Valley to Chur, along the Inn nearly to
Kufstein, and for some distance along the
Enns. Thus, then, five great rivers have
taken advantage of this main fold, each of

them eventually breaking through into a transverse valley.

The Pusterthal in the Tyrol offers us an interesting case of what is obviously a single valley, which has, however, been slightly raised in the centre, near Toblach, so that from this point the water flows in opposite directions — the Drau eastward, and the Rienz westward. In this case the elevation is single and slight: in the main valley there are several, and they are much loftier, still we may, I think, regard that of the Isère from Chambery to Albertville, of the Rhone from Martigny to its source, of the Urseren Thal, of the Vorder Rhine from its source to Chur, of the Inn from Landeck to below Innsbruck, even perhaps of the Enns from Radstadt to Hieflau as in one sense a single valley, due to one of these longitudinal folds, but interrupted by bosses of gneiss and granite, — one culminating in Mont Blanc, and another in the St. Gotthard, — which have separated the waters of the Isère, the Rhone, the Vorder Rhine, the Inn, and the Enns. That the valley of

Chamouni, the Valais, the Urseren Thal, and that of the Vorder Rhine really form part of one great fold is further shown by the presence of a belt of Jurassic strata nipped in, as it were, between the crystalline rocks.

This seems to throw light on the remarkable turns taken by the Rhone at Martigny and the Vorder Rhine at Chur, where they respectively quit the great longitudinal fold, and fall into secondary transverse valleys. The Rhone for the upper part of its course, as far as Martigny, runs in the great longitudinal fold of the Valais; at Martigny it falls into and adopts the transverse valley, which properly belongs to the Dranse; for the Dranse is probably an older river and ran in the present course even before the great fold of the Valais. This would seem to indicate that the Oberland range is not so old as the Pennine, and that its elevation was so gradual that the Dranse was able to wear away a passage as the ridge gradually rose. After leaving the Lake of Geneva the Rhone follows a course curving gradually to the

south, until it reaches St. Genix, where it falls
into and adopts a transverse valley which
properly belongs to the little river Guiers; it
subsequently joins the Ain and finally falls
into the Saône. If these valleys were attributed to their older occupiers we should therefore confine the name of the Rhone to the
portion of its course from the Rhone glacier to
Martigny. From Martigny it occupies successively the valleys of the Dranse, Guiers, Ain,
and Saône. In fact, the Saône receives the
Ain, the Ain the Guiers, the Guiers the
Dranse, and the Dranse the Rhone. This is
not a mere question of names, but also one of
antiquity. The Saône, for instance, flowed
past Lyons to the Mediterranean for ages
before it was joined by the Rhone. In our
nomenclature, however, the Rhone has swal-
lowed up the others. This is the more curious
because of the three great rivers which unite
to form the lower Rhone, namely, the Saône,
the Doubs, and the Rhone itself, the Saône
brings for a large part of the year the
greatest volume of water, and the Doubs
has the longest course. Other similar cases

might be mentioned.   The Aar, for instance,
is a somewhat larger river than the Rhine.

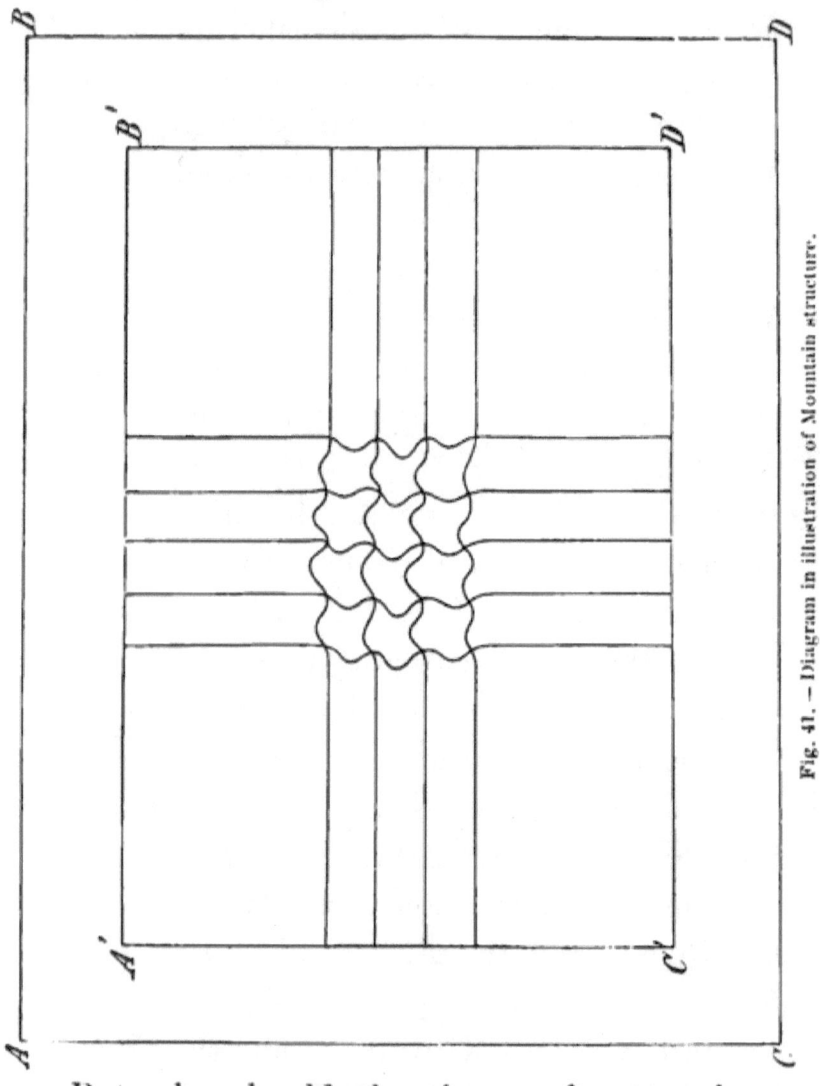

Fig. 41. — Diagram in illustration of Mountain structure.

But why should the rivers, after running

for a certain distance in the direction of the
main axis. so often break away into lateral
valleys? If the elevation of a chain of mountains be due to the causes suggested in p. 214,
it is evident, though, so far as I am aware,
stress has not hitherto been laid upon this,
that the compression and consequent folding
of the strata (Fig. 41) would not be in the
direction A B only, but also at right angles to
it, in the direction A C, though the amount of
folding might be much greater in one direction than in the other. Thus in the case of
Switzerland. while the main folds run south-
west by north-east, there would be others at
right angles to the main axis. The complex
structure of the Swiss mountains may be
partly due to the coexistence of these two
directions of pressure at right angles to one
another. The presence of a fold so originating
would often divert the river to a course more
or less nearly at right angles to its original
direction.

Switzerland, moreover, slopes northwards
from the Alps. so that the lowest part of the
great Swiss plain is that along the foot of

the Jura. Hence the main drainage runs along the line from Yverdun to Neuchâtel, down the Zihl to Soleure, and then along the Aar to Waldshut: the Upper Aar, the Emmen, the Wiggern, the Suhr, the Wynen, the lower Reuss, the Sihl, and the Limmat, besides several smaller streams, running approximately parallel to one another north-north-east, and at angles to the main axis of elevation, and all joining the Aar from the south, while on the north it does not receive a single contributary of any importance.

On the south side of the Alps again we have the Dora Baltea, the Sesia, the Ticino, the Olonna, the Adda, the Adige, etc., all running south-south-east from the axis of elevation to the Po.

Indeed, the general slope of Switzerland, being from the ridge of the Alps towards the north, it will be observed (Fig. 42) that almost all the large affluents of these rivers running in longitudinal valleys fall in on the south, as, for instance, those of the Isère from Albertville to Grenoble, of the Rhone from its source to Martigny, of the Vorder Rhine from its source to Chur, of the Inn from Landeck to Kufstein,

Fig. 42.

of the Enns from its source to near Admont,

of the Danube from its source to Vienna, and as just mentioned, of the Aar from Bern to Waldshut. Hence also, whenever the Swiss rivers running east and west break into a transverse valley, as the larger ones all do, and some more than once, they invariably, whether originally running east or westwards, turn towards the north.

But although we thus get a clue to the general structure of Switzerland, the whole question is extremely complex, and the strata have been crumpled and folded in the most complicated manner, sometimes completely reversed, so that older rocks have been folded back on younger strata, and even in some cases these folds again refolded. Moreover, the denudation by aerial action, by glaciers, frosts, and rivers has removed hundreds, or rather thousands, of feet of strata. In fact, the mountain tops are not by any means the spots which have been most elevated, but those which have been least denuded; and hence it is that so many of the peaks stand at about the same altitude.

THE CONFLICTS AND ADVENTURES OF RIVERS

Our ancestors looked upon rivers as being in some sense alive, and in fact in their "struggle for existence" they not only labour to adapt their channel to their own requirements, but in many cases enter into conflict with one another.

In the plain of Bengal, for instance, there are three great rivers, the Brahmapootra coming from the north, the Ganges from the west, and the Megna from the east, each of them with a number of tributary streams. Mr. Fergusson[1] has given us a most interesting and entertaining account of the struggles between these great rivers to occupy the fertile plain of Bengal.

The Megna, though much inferior in size to the Brahmapootra, has one great advantage. It depends mainly on the monsoon rains for its supply, while the Brahmapootra not only has a longer course to run, but relies for its floods, to a great extent, on the melting of the

[1] *Geol. Jour.*, 1863.

snow, so that, arriving later at the scene of the struggle. it finds the country already occupied by the Megna to such an extent that it has been driven nearly 70 miles northwards, and forced to find a new channel.

Under these circumstances it has attacked the territory of the Ganges, and being in flood earlier than that river, though later than the Megna, it has in its turn a great advantage.

Whatever the ultimate result may be the struggle continues vigorously. At Sooksaghur, says Fergusson, " there was a noble country house, built by Warren Hastings, about a mile from the banks of the Hoogly. When I first knew it in 1830 half the avenue of noble trees, which led from the river to the house. was gone ; when I last saw it, some eight years afterwards, the river was close at hand. Since then house. stables, garden, and village are all gone, and the river was on the point of breaking through the narrow neck of high land that remained, and pouring itself into some weak-banded nullahs in the lowlands beyond : and if it had succeeded, the Hoogly would have deserted Calcutta. At this juncture the Eastern Bengal Railway Company intervened. They were carrying their works along the ridge, and they have, for the moment at least, stopped the oscillation in this direction."

This has affected many of the other tributaries of the Ganges, so that the survey made by Rennell in 1780–90 is no longer any evidence as to the present course of the rivers. They may now be anywhere else; in some cases all we can say is that they are certainly not now where they were then.

The association of the three great European rivers, the Rhine, the Rhone, and the Danube, with the past history of our race, invests them with a singular fascination, and their past history is one of much interest. They all three rise in the group of mountains between the Galenstock and the Bernardino, within a space of a few miles; on the east the waters run into the Black Sea, on the north into the German Ocean, and on the west into the Mediterranean. But it has not always been so. Their head-waters have been at one time interwoven together.

At present the waters of the Valais escape
from the Lake of Geneva at the western end,
and through the remarkable defile of Fort de
l'Ecluse and Malpertius, which has a depth of
600 feet, and is at one place not more than
14 feet across. Moreover, at various points
round the Lake of Geneva, remains of lake
terraces show that the water once stood at a
level much higher than the present. One
of these is rather more than 250 feet [1] above
the lake.

A glance at the map will show that between Lausanne and Yverdun there is a low
tract of land, and the Venoge, which falls
into the Lake of Geneva between Lausanne
and Morges, runs within about half a mile of
the Nozon, which falls into the Lake of Neuchâtel at Yverdun, the two being connected
by the Canal d'Entreroches, and the height
of the watershed being only 76 metres (250
feet), corresponding with the above mentioned
lake terrace. It is evident, therefore, that
when the Lake of Geneva stood at the level of
the 250 feet terrace the waters ran out, not as

[1] Favre, *Rech. Geol. de la Savoie.*

now at Geneva and by Lyons to the Mediter-
ranean, but near Lausanne by Cissonay and
Entreroches to Yverdun, and through the
Lake of Neuchâtel into the Aar and the Rhine.

But this is not the whole of the curious
history. At present the Aar makes a sharp
turn to the west at Waldshut, where it falls
into the Rhine, but there is reason to believe
that at a former period, before the Rhine had
excavated its present bed, the Aar continued
its course eastward to the Lake of Constance,
by the valley of the Klettgau, as is indicated by
the presence of gravel beds containing pebbles
which have been brought, not by the Rhine
from the Grisons, but by the Aar from the
Bernese Oberland, showing that the river
which occupied the valley was not the Rhine
but the Aar. It would seem also that at an
early period the Lake of Constance stood at a
considerably higher level, and that the outlet
was, perhaps, from Frederichshaven to Ulm,
along what are now the valleys of the
Schussen and the Ried, into the Danube.

Thus the head-waters of the Rhone appear
to have originally run by Lausanne and the

Lake of Constance into the Danube, and so to
the Black Sea. Then, after the present valley
was opened between Waldshut and Basle,
they flowed by Basle and the present Rhine,
and after joining the Thames, over the plain
which now forms the German Sea into the
Arctic Ocean between Scotland and Norway.
Finally, after the opening of the passage at
Fort de l'Ecluse, by Geneva, Lyons, and the
Valley of the Saône, to the Mediterranean.

It must not, however, be supposed that
these changes in river courses are confined to
the lower districts. Mountain streams have
also their adventures and vicissitudes, their
wars and invasions. Take for instance the
Upper Rhine, of which we have a very interesting account by Heim. It is formed of
three main branches, the Vorder Rhine, Hinter
Rhine, and the Albula. The two latter, after
meeting near Thusis, unite with the Vorder
Rhine at Reichenau, and run by Chur, Mayenfeld, and Sargans into the Lake of Constance at Rheineck. At some former period,
however, the drainage of this district was
very different, as is shown in Fig. 43.

The Vorder and Hinter Rhine united then (Fig. 43) as they do now at Reichenau, but at a much higher level, and ran to Mayenfeld, not by Chur, but by the Kunckel Pass to Sargans, and so on, not to the Lake of Constance, but to that of Zurich. The Landwasser at that time rose in the Schlappina Joch, and after receiving as tributaries the Vereina and the Sardasca, joined the Albula, as it does now at Tiefenkasten; but instead of going round to meet the Hinter Rhine near Thusis, the two together travelled parallel with, but at some distance from, the Hinter Rhine, by Heide to Chur, and so to Mayenfeld.

In the meanwhile, however, the Landquart was stealthily creeping up the valley, attacked the ridge which then united the Casanna and the Madrishorn, and gradually forcing the passage, invaded (Fig. 44) the valleys of the Schlappina, Vereina, and Sardasca, absorbed them as tributaries, and, detaching them from their allegiance to the Landwasser, annexed the whole of the upper province which had formerly belonged to that river.

The Schyn also gradually worked its way upwards from Thusis till it succeeded in sapping the Albula, and carried it down the

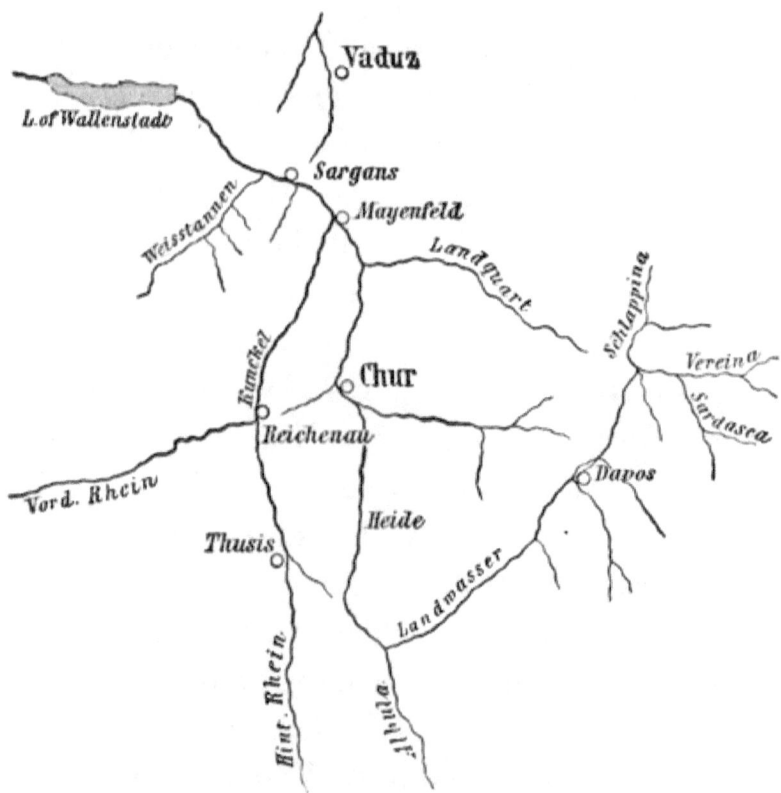

Fig. 43. — River system round Chur, as it used to be.

valley to join the Vorder Rhine near Thusis. In what is now the main valley of the Rhine above Chur another stream ate its way back, and eventually tapped the main river at Reichenau, thus diverting it from the Kunckel, and carrying it round by Chur.

At Sargans a somewhat similar process

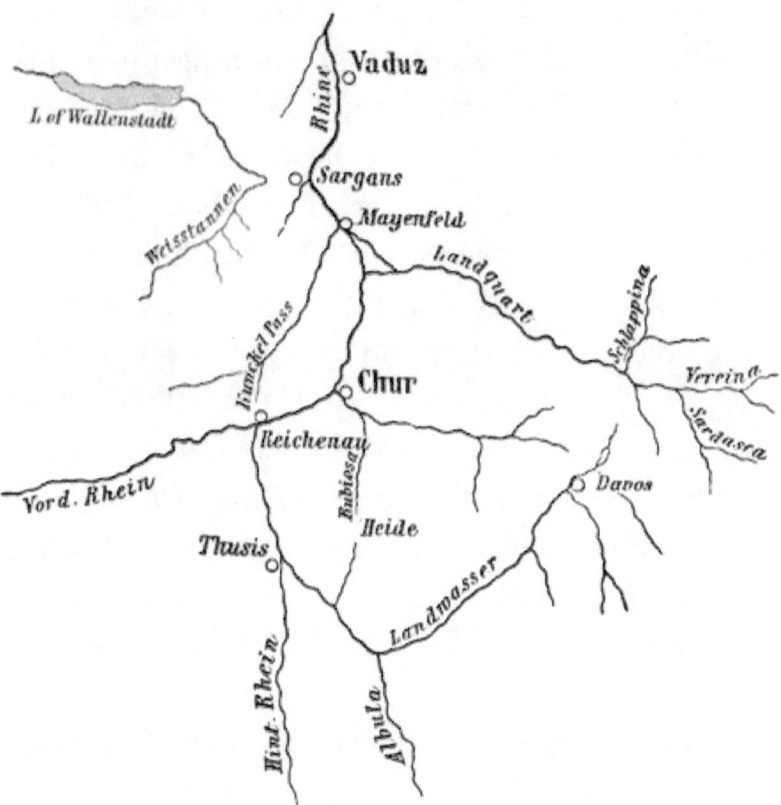

Fig. 44. — River system round Chur, as it is.

was repeated, with the addition that the material brought down by the Weisstannen, or perhaps a rockfall, deflected the Rhine, just as we see in Fig. 30 that the Rhone was pushed on one side by the Borgne. The Rhone, however, had no choice, it was obliged to force, and has forced its way over the cone deposited by the Borgne. The Rhine, on the contrary, had the option of running down by Vaduz to Rheinach, and has adopted this course. The watershed between it and the Weisstannen is, however, only about 20 feet in height, and the people of Zurich watch it carefully, lest any slight change should enable the river to return to its old bed. The result of all these changes is that the rivers have changed their courses from those shown in Fig. 43 to their present beds as shown in Fig. 44.

Another interesting case is that of the Upper Engadine (Fig. 45), to which attention has been called by Bonney and Heim. The fall of the Val Bregaglia is much steeper than that of the Inn, and the Maira has carried off the head-waters of that river away into Italy. The Col was formerly perhaps as far south as Stampa: the Albegna, the Upper Maira, and the stream from the Forgno Glacier, originally belonged to the Inn, but have been captured by the Lower Maira. Their direction still indicates this; they seem as if they regretted the unwelcome change, and yearned to rejoin their old companions.

Moreover, as rivers are continually cutting back their valleys they must of course sometimes meet. In these cases when the valleys are at different levels the lower rivers have drained the upper ones, and left dry, deserted valleys. In other cases, especially in flatter districts, we have bifurcations, as, for instance, at Sargans, and several of the Italian lakes. Every one must have been struck by the peculiar bifurcation of the Lakes of Como and Lugano, while a very slight depression would connect the Lake Varese with the Maggiore, and give it also a double southern end.

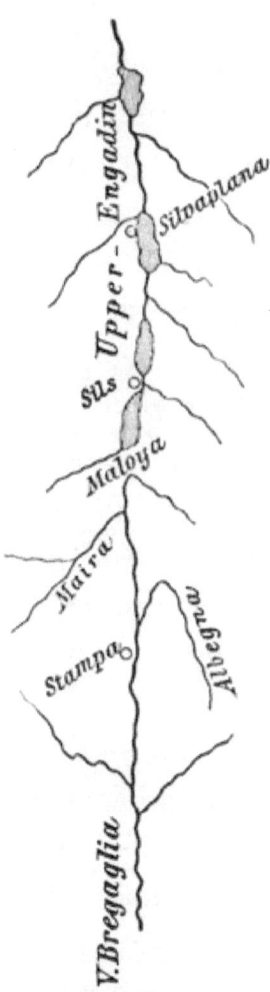

Fig. 45.—River system of the Maloya.

ON LAKES

The problem of the origin of Lakes is by no means identical with that of Valleys. The latter are due, primarily as a rule to geological causes, but so far as their present condition is concerned, mainly to the action of rain and rivers. Flowing water, however, cannot give rise to lakes.

It is of course possible to have valleys without lakes, and in fact the latter are, now at least, exceptional. There can be no lakes if the slope of the valley is uniform. To what then are lakes due?

Professor Ramsay divides Lakes into three classes:—

1. Those due to irregular accumulations of drift, and which are generally quite shallow.

2. Those formed by moraines.

3. Those which occupy true basins scooped by glacier ice out of the solid rock.

To these must, however, I think be added at least one other great class and several minor ones, namely,—

4. Those due to inequalities of elevation or depression.

5. Lakes in craters of extinct volcanoes, for instance, Lake Avernus.

6. Those caused by subsidence due to the removal of underlying soluble rocks, such as some of the Cheshire Meres.

7. Loop lakes in deserted river courses, of which there are many along the course of the Rhine.

8. Those due to rockfalls, landslips, or lava currents, damming up the course of a river.

9. Those caused by the advance of a glacier across a lateral valley, such as the Mergelen See, or the ancient lake whose margins form the celebrated " Parallel Roads of Glen Roy."

As regards the first class we find here and there on the earth's surface districts sprinkled with innumerable shallow lakes of all sizes, down to mere pools. Such, for instance, occur in the district of Le Doubs between the Rhone and the Saône, that of La Sologne near Orleans, in parts of North America, and in Finland. Such lakes are, as a rule, quite shallow. Some geologists, Geikie, for instance, ascribe them to the fact of these regions having been covered by sheets of ice which strewed the land with irregular masses of clay, gravel, and sand, lying on a stratum impervious to water, either of hard rock such as granite or gneiss, or of clay, so that the rain cannot percolate through it, and without sufficient inclination to throw it off.

2. To Ramsay's second class of Lakes belong those formed by moraines. The materials forming moraines being, however, comparatively loose, are easily cut through by streams. There are in Switzerland many cases of valleys crossed by old moraines, but they have generally been long ago worn through by the rivers.

3. Ramsay and Tyndall attribute most of the great Swiss and Italian lakes to the action of glaciers, and regard them as rock basins. It is of course obvious that rivers cannot make basin-shaped hollows surrounded by rock on all sides. The Lake of Geneva, 1230 feet above the sea, is over 1000 feet deep; the Lake of Brienz is 1850 feet above

the sea, and 2000 feet deep, so that its
bottom is really below the sea level. The
Italian Lakes are even more remarkable.
The Lake of Como, 700 feet above the sea,
is 1929 feet deep. Lago Maggiore, 685 feet
above the sea, is no less than 2625 feet
deep.

If the mind is at first staggered at the
magnitude of the scale, we must remember
that the ice which is supposed to have scooped
out the valley in which the Lake of Geneva
now reposes, was once at least 4000 feet
thick; while the moraines were also of
gigantic magnitude, that of Ivrea, for instance, being no less than 1500 feet above
the river, and several miles long.

Indeed it is obvious that a glacier many
hundred, or in some cases several thousand,
feet in thickness, must exercise great pressure
on the bed over which it travels. We see
this from the striæ and grooves on the solid
rocks, and the fine mud which is carried down
by glacial streams. The deposit of glacial
rivers, the "loess" of the Rhine itself, is
mainly the result of this ice-waste, and that is

why it is so fine, so impalpable. That glaciers do deepen their beds seems therefore unquestionable.

Moreover, though the depth of some of these lakes is great, the true slope is very slight.

Tyndall and Ramsay do not deny that the original direction of valleys, and consequently of lakes, is due to cosmical causes and geological structure, while even those who have most strenuously opposed the theory which attributes lakes to glacial erosion do not altogether deny the action of glaciers. Favre himself admits that " it is impossible to deny that valleys, after their formation, have been swept out and perhaps enlarged by rivers and glaciers."

Even Ruskin admits " that a glacier may be considered as a vast instrument of friction, a white sand-paper applied slowly but irresistibly to all the roughness of the hill which it covers."

It is obvious that sand-paper applied " irresistibly" and long enough, must gradually wear away and lower the surface.

I cannot therefore resist the conclusion that glaciers have taken an important part in the formation of lakes.

The question has sometimes been discussed as if the point at issue were whether rivers or glaciers were the most effective as excavators. But this is not so. Those who believe that lakes are in many cases due to glaciers might yet admit that rivers have greater power of erosion. There is, however, an essential difference in the mode of action. Rivers tend to regularise their beds; they drain, rather than form lakes. Their tendency is to cut through any projections so that finally their course assumes some such curve as that below, from the source (a) to its entrance into the sea (b).

Fig. 46. — Final Slope of a River.

Glaciers, however, have in addition a scooping power, so that if similarly a d b in Fig. 47 represent the course of a glacier, starting at a and gradually thinning out to c, it may scoop out the rock to a certain extent at d; in that case if it subsequently retires say to c, there would be a lake lying in the basin thus formed between c and e.

Fig. 47.

On the other hand I am not disposed to attribute the Swiss lakes altogether to the action of glaciers. In the first place it does not seem clear that they occupy true rock basins. On this point more evidence is required. That some lakes are due to unequal changes of level will hardly be denied. No one, for instance, as Bonney justly observes,[1] would attribute the Dead Sea to glacial erosion.

The Alps, as we have seen, are a succession of great folds, and there is reason to regard the central one as the oldest. If then the same process continued, and the outer fold was still further raised, or a new one formed, more quickly than the rivers could cut it

[1] *Growth and Structure of the Alps.*

back, they would be dammed up, and lakes would result.

Moreover, if the formation of a mountain region be due to subsidence, and consequent crumpling, as indicated on p. 217, so that the strata which originally occupied the area A B C D are compressed into A′ B′ C′ D′, it is evident, as already mentioned, that while the line of least resistance, and, consequently, the principal folds might be in the direction A′ B′, there must also be a tendency to the formation of similar folds at right angles, or in the direction A′ C′. Thus, in the case of Switzerland, while the main folds run south-west by north-east there would also be others at right angles, though the amount of folding might be much greater in the one direction than in the other. To this cause the bosses, for instance — at Martigny, the Furca, and the Ober Alp, — which intersect the great longitudinal valley of Switzerland, are perhaps due.

The great American lakes also are probably due to differences of elevation. Round Lake Ontario, for instance, there is a raised beach which at the western end of the lake is 363 feet above the sea level, but rises towards the East and North until near Fine it reaches an elevation of 972 feet. As this terrace must have been originally horizontal we have here a lake barrier, due to a difference of elevation, amounting to over 600 feet.

In the same way we get a clue to the curious cruciform shape of the Lake of Lucerne as contrasted with the simple outline of such lakes as those of Neuchâtel or Zurich. That of Lucerne is a complex lake. Soundings have shown that the bottom of the Urner See is quite flat. It is in fact the old bed of the Reuss, which originally ran, not as now by Lucerne, but by Schwytz and through the Lake of Zug. In the same way the Alpnach See is the old bed of the Aa, which likewise ran through the Lake of Zug. The old river terraces of the Reuss can be traced in places between Brunnen and Goldau. Now these terraces must have originally sloped from the upper part downwards, from Brunnen towards Goldau. But at present the slope is the other way, i.e. from Goldau towards Brunnen. From this and other evidence we conclude

that in the direction from Lucerne towards
Rapperschwyl there has been an elevation of
the land, which has dammed up the valleys
and thus turned parts of the Aa and the
Reuss into lakes — the two branches of the
Lake of Lucerne known as the Alpnach See
and Urner See.

During the earthquakes of 1819 while part
of the Runn of Cutch, 2000 square miles in
area, sunk several feet, a ridge of land, called
by the natives the Ulla-Bund or "the wall of
God," thirty miles long, and in parts sixteen
miles wide, was raised across an ancient arm
of the Indus, and turned it temporarily into
a lake.

In considering the great Italian lakes,
which descend far below the sea level, we
must remember that the Valley of the Po is a
continuation of the Adriatic, now filled up
and converted into land, by the materials
brought down from the Alps. Hence we are
tempted to ask whether the lakes may not
be remains of the ancient sea which once
occupied the whole plain. Moreover just as
the Seals of Lake Baikal in Siberia carry us

back to the time when that great sheet of fresh water was in connection with the Arctic Ocean, so there is in the character of the Fauna of the Italian lakes, and especially the presence of a Crab in the Lake of Garda, some confirmation of such an idea. Further evidence, however, is necessary before these interesting questions can be definitely answered.

Lastly, some lakes and inland seas seem to be due to even greater cosmical causes. Thus a line inclined ten degrees to the pole beginning at Gibraltar would pass through a great chain of inland waters — the Mediterranean, Black Sea, Caspian, Aral, Baikal, and back again through the great American lakes.

But though many causes have contributed to the original formation and direction of Valleys, their present condition is mainly due to the action of water. When we contemplate such a valley, for example, as that which is called *par excellence* the " Valais," we can at first hardly bring ourselves to realise this; but we can trace up valleys, from the little

watercourse made by last night's rains up to
the greatest valleys of all.

These considerations, however, do not of
course apply to such depressions as those
of the great oceans. These were probably
formed when the surface of the globe began
to solidify, and, though with many modifications, have maintained their main features
ever since.

ON THE CONFIGURATION OF VALLEYS

The conditions thus briefly described repeat
themselves in river after river, valley after
valley, and it adds, I think, very much to the
interest with which we regard them if, by
studying the general causes to which they are
due, we can explain their origin, and thus to
some extent understand the story they have
to tell us, and the history they record.

What, then, has that history been? The
same valley may be of a very different character, and due to very different causes, in different parts of its course. Some valleys are
due to folds (see Fig. 41) caused by subterra-

nean changes, but by far the greater number
are, in their present features, mainly the result of erosion.   As soon as any tract of land
rose out of the sea, the rain which fell on the
surface would trickle downwards in a thousand rills, forming pools here and there (see
Fig. 37), and gradually collecting into larger
and larger streams.   Wherever the slope was
sufficient the water would begin cutting into
the soil and carrying it off to the sea.   This
action would be the same in any case, but,
of course, would differ in rapidity according
to the hardness of the ground.   On the
other hand, the character of the valley
would depend greatly on the character of
the strata, being narrow where they were
hard and tough ; broader, on the contrary,
where they were soft, so that they crumbled
readily into the stream, or where they were
easily split by the weather.   Gradually the
stream would eat into its bed until it reached
a certain slope, the steepness of which would
depend on the volume of water.   The erosive
action would then cease, but the weathering
of the sides and consequent widening would

continue, and the river would wander from one part of its valley to another, spreading the materials and forming a river plain. At length, as the rapidity still further diminished, it would no longer have sufficient power even to carry off the materials brought down. It would form, therefore, a cone or delta, and instead of meandering, would tend to divide into different branches. These three stages, we may call those of —

1. Deepening and widening;
2. Widening and levelling;
3. Filling up;

and every place in the second stage has passed through the first; every one in the third has passed through the second.

A velocity of 6 inches per second will lift fine sand, 8 inches will move sand as coarse as linseed, 12 inches will sweep along fine gravel, 24 inches will roll along rounded pebbles an inch diameter, and it requires 3 feet per second at the bottom to sweep along angular stones of the size of an egg.

When a river has so adjusted its slope that it neither deepens its bed in the upper portion

of its course, nor deposits materials, it is said
to have acquired its "regimen," and in such
a case if the character of the soil remains the
same, the velocity must also be uniform. The
enlargement of the bed of a river is not, however, in proportion to the increase of its waters as it approaches the sea. If, therefore,
the slope did not diminish, the regimen would
be destroyed, and the river would again com-
mence to eat out its bed. Hence as rivers
enlarge, the slope diminishes, and consequently
every river tends to assume some such "regimen" as that shown in Fig. 46.

Now, suppose that the fall of the river is
again increased, either by a fresh elevation,
or locally by the removal of a barrier. Then
once more the river regains its energy. Again
it cuts into its old bed, deepening the valley,
and leaving the old plain as a terrace high
above its new course. In many valleys several such terraces may be seen, one above
the other. In the case of a river running in a
transverse valley, that is to say of a valley
lying at right angles to the "strike" or direction of the strata (such, for instance, as the

Reuss), the water acts more effectively than
in longitudinal valleys running along the
strike.   Hence the lateral valleys have been
less deeply excavated than that of the Reuss
itself, and the streams from them enter the
main valley by rapids or cascades.   Again,
rivers running in transverse valleys cross
rocks which in many cases differ in hardness,
and of course they cut down the softer strata
more rapidly than the harder ones; each ridge
of harder rock will therefore form a dam and
give rise to a rapid, or cataract.   We often
as we ascend a river, after a comparatively
flat plain, find ourselves in a narrow defile,
down which the water rushes in an impet-
uous torrent, but at the summit of which,
to our surprise, we find another broad flat
valley.

Another lesson which we learn from the
study of river valleys, is that, just as geological
structure was shown by Sir C. Lyell to be no
evidence of cataclysms, but the result of slow
action; so also the excavation of valleys is
due mainly to the regular flow of rivers ; and
floods, though their effects are more sudden

and striking, have had, after all, comparatively little part in the result.

The mouths of rivers fall into two principal classes. If we look at any map we cannot but be struck by the fact that some rivers terminate in a delta, some in an estuary. The Thames, for instance, ends in a noble estuary, to which London owes much of its wealth and power. It is obvious that the Thames could not have excavated this estuary while the coast was at its present level. But we know that formerly the land stood higher, that the German Ocean was once dry land, and the Thames, after joining the Rhine, ran northwards, and fell eventually into the Arctic Ocean. The estuary of the Thames, then, dates back to a period when the south-east of England stood at a higher level than the present, and even now the ancient course of the river can be traced by soundings under what is now sea. The sites of present deltas, say of the Nile, were also once under water, and have been gradually reclaimed by the deposits of the river.

It would indeed be a great mistake to

suppose that rivers always tend to deepen
their valleys. This is only the case when the
slope exceeds a certain angle. When the fall
is but slight they tend on the contrary to
raise their beds by depositing sand and mud
brought down from higher levels. Hence in
the lower part of their course many of the
most celebrated rivers — the Nile. the Po. the
Mississippi. the Thames, etc. — run upon em-
bankments, partly of their own creation.

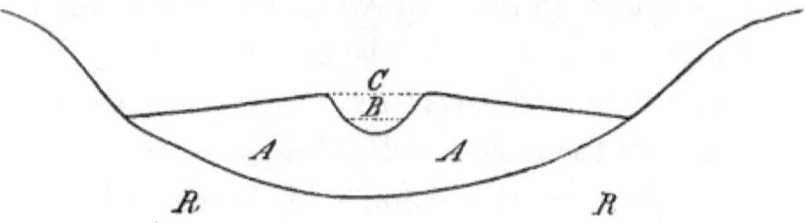

Fig. 48. — Diagrammatic section of a valley (exaggerated)

R R, rocky basis of valley; *A A*, sedimentary strata; *B*, ordinary level
of river; *C*, flood level.

The Reno, the most dangerous of all the
Apennine rivers. is in some places as much as
30 feet above the adjoining country. Rivers
under such conditions. when not interfered
with by Man. sooner or later break through
their banks. and leaving their former bed,
take a new course along the lowest part of

their valley, which again they gradually raise above the rest. Hence, unless they are kept in their own channels by human agency, such rivers are continually changing their course.

If we imagine a river running down a regularly inclined plane in a more or less straight line; any inequality or obstruction would produce an oscillation, which when once started would go on increasing until the force of gravity drawing the water in a straight line downwards equals that of the force tending to divert its course. Hence the radius of the curves will follow a regular law depending on the volume of water and the angle of inclination of the bed. If the fall is 10 feet per mile and the soil homogeneous, the curves would be so much extended that the course would appear almost straight. With a fall of 1 foot per mile the length of the curve is, according to Fergusson, about six times the width of the river, so that a river 1000 feet wide would oscillate once in 6000 feet. This is an important considera-tion, and much labour has been lost in trying

to prevent rivers from following their natural
law of oscillation.   But rivers are very true to
their own laws, and a change at any part is
continued both upwards and downwards, so
that a new oscillation in any place cuts its
way through the whole plain of the river both
above and below.

The curves of the Mississippi are, for instance, for a considerable part of its course
so regular that they are said to have been
used by the Indians as a measure of distance.

If the country is flat a river gradually
raises the level on each side, the water which
overflows during floods being retarded by
reeds, bushes, trees, and a thousand other
obstacles, gradually deposits the solid matter
which it contains, and thus raising the surface, becomes at length suspended, as it were,
above the general level.   When this elevation
has reached a certain point, the river during
some flood bursts its banks, and deserting its
old bed takes a new course along the lowest
accessible level.   This then it gradually fills
up, and so on ; coming back from time to

time if permitted, after a long cycle of years, to its first course.

In evidence of the vast quantity of sediment which rivers deposit, I may mention that the river-deposits at Calcutta are more than 400 feet in thickness.

In addition to temporary " spates," due to heavy rain, most rivers are fuller at one time of year than another, our rivers, for instance, in winter, those of Switzerland, from the melting of the snow, in summer. The Nile commences to rise towards the beginning of July ; from August to October it floods all the low lands, and early in November it sinks again. At its greatest height the volume of water sometimes reaches twenty times that when it is lowest, and yet perhaps not a drop of rain may have fallen. Though we now know that this annual variation is due to the melting of the snow and the fall of rain on the high lands of Central Africa, still when we consider that the phenomenon has been repeated annually for thousands of years it is impossible not to regard it with wonder. In fact Egypt itself may be said to be the bed of the Nile in flood time.

Some rivers, on the other hand, offer no
such periodical differences. The lower Rhone,
for instance, below the junction with the
Saône, is nearly equal all through the year,
and yet we know that the upper portion is
greatly derived from the melting of the Swiss
snows. In this case, however, while the
Rhone itself is on this account highest in
summer and lowest in winter, the Saône, on
the contrary, is swollen by the winter's rain,
and falls during the fine weather of summer.
Hence the two tend to counterbalance one
another.

Periodical differences are of course comparatively easy to deal with. It is very different with floods due to irregular rainfall.
Here also, however, the mere quantity of rain
is by no means the only matter to be considered. For instance a heavy rain in the
watershed of the Seine, unless very prolonged,
causes less difference in the flow of the river,
say at Paris, than might at first have been
expected, because the height of the flood in
the nearer affluents has passed down the river
before that from the more distant streams has

arrived. The highest level is reached when the rain in the districts drained by the various affluents happens to be so timed that the different floods coincide in their arrival at Paris.

CHAPTER IX

THE SEA

There is a pleasure in the pathless woods,
There is a rapture on the lonely shore,
There is society, where none intrudes,
By the deep Sea, and music in its roar:
I love not Man the less, but Nature more,
From these our interviews, in which I steal
From all I may be, or have been before,
To mingle with the Universe, and feel
What I can ne'er express, yet cannot all conceal.

Roll on, thou deep and dark-blue Ocean — roll!

BYRON.

THE LAND'S END.

CHAPTER IX

THE SEA

WHEN the glorious summer weather comes, when we feel that by a year's honest work we have fairly won the prize of a good holiday, how we turn instinctively to the Sea. We pine for the delicious smell of the sea air, the murmur of the waves, the rushing sound of the pebbles on the sloping shore, the cries of the sea-birds; and long to

> Linger, where the pebble-paven shore,
> Under the quick, faint kisses of the Sea,
> Trembles and sparkles as with ecstasy.[1]

How beautiful the sea-coast is! At the foot of a cliff, perhaps of pure white chalk, or rich red sandstone, or stern grey granite, lies the shore of gravel or sand, with a few

---

[1] Shelley.

scattered plants of blue Sea Holly, or yellow-
flowered Horned Poppies, Sea-kale, Sea Convolvulus, Saltwort, Artemisia, and Sea-grasses;
the waves roll leisurely in one by one, and as
they reach the beach, each in turn rises up in
an arch of clear, cool, transparent, green
water, tipped with white or faintly pinkish
foam, and breaks lovingly on the sands;
while beyond lies the open Sea sparkling in
the sunshine.

> . . . O pleasant Sea
> Earth hath not a plain
> So boundless or so beautiful as thine.[1]

The Sea is indeed at times overpoweringly
beautiful. At morning and evening a sheet
of living silver or gold, at mid-day deep blue;
even

> Too deeply blue; too beautiful; too bright;
> Oh, that the shadow of a cloud might rest
> Somewhere upon the splendour of thy breast
> In momentary gloom.[2]

There are few prettier sights than the beach
at a seaside town on a fine summer's day;
the waves sparkling in the sunshine, the water

[1] Campbell.   [2] Holmes.

and sky each bluer than the other, while the
sea seems as if it had nothing to do but to
laugh and play with the children on the sands;
the children perseveringly making castles with
spades and pails, which the waves then run
up to and wash away, over and over and
over again, until evening comes and the children go home, when the Sea makes everything smooth and ready for the next day's
play.

Many are satisfied to admire the Sea from
shore, others more ambitious or more free
prefer a cruise. They feel with Tennyson's
voyager:

> We left behind the painted buoy
>   That tosses at the harbour-mouth;
> And madly danced our hearts with joy,
>   As fast we fleeted to the South:
> How fresh was every sight and sound
>   On open main or winding shore!
> We knew the merry world was round,
>   And we might sail for evermore.

Many appreciate both. The long roll of
the Mediterranean on a fine day (and I suppose even more of the Atlantic, which I have
never enjoyed), far from land in a good ship,

and with kind friends, is a joy never to be forgotten.

To the Gulf Stream and the Atlantic Ocean Northern Europe owes its mild climate. The same latitudes on the other side of the Atlantic are much colder. To find the same average temperature in the United States we must go far to the south. Immediately opposite us lies Labrador, with an average temperature the same as that of Greenland; a coast almost destitute of vegetation. a country of snow and ice, whose principal wealth consists in its furs, and a scattered population, mainly composed of Indians and Esquimaux. But the Atlantic would not alone produce so great an effect. We owe our mild and genial climate mainly to the Gulf Stream — a river in the ocean, twenty million times as great as the Rhone — the greatest, and for us the most important, river in the world, which brings to our shores the sunshine of the West Indies.

The Sea is outside time. A thousand, ten thousand, or a million years ago it must have looked just as it does now, and as it will ages hence. With the land this is not so. The mountains and hills, rivers and valleys, animals and plants are continually changing: but the Sea is always the same,

> Steadfast, serene, immovable, the same
> Year after year.

Directly we see the coast, or even a ship, the case is altered. Boats may remain the same for centuries, but ships are continually being changed. The wooden walls of old England are things of the past, and the iron-clads of to-day will soon be themselves improved off the face of the ocean.

The great characteristic of Lakes is peace, that of the Sea is energy, somewhat restless, perhaps, but still movement without fatigue.

> The Earth lies quiet like a child asleep,
> The deep heart of the Heaven is calm and still,
> Must thou alone a restless vigil keep,
> And with thy sobbing all the silence fill.[1]

A Lake in a storm rather gives us the impression of a beautiful Water Spirit tormented by some Evil Demon; but a storm at Sea is one of the grandest manifestations of Nature.

[1] Bell.

Yet more ; the billows and the depths have more ;
  High hearts and brave are gathered to thy breast ;
They hear not now the booming waters roar,
  The battle thunders will not break their rest.
Keep thy red gold and gems, thou stormy grave ;
    Give back the true and brave.[1]

The most vivid description of a storm at sea is, I think, the following passage from Ruskin's *Modern Painters :*

"Few people, comparatively, have ever seen the effect on the sea of a powerful gale continued without intermission for three or four days and nights ; and to those who have not, I believe it must be unimaginable, not from the mere force or size of the surge, but from the complete annihilation of the limit between sea and air. The water from its prolonged agitation is beaten, not into mere creaming foam, but into masses of accumulated yeast, which hangs in ropes and wreaths from wave to wave, and, where one curls over to break, form a festoon like a drapery from its edge ; these are taken up by the wind, not in dissipating dust, but bodily, in writhing, hanging, coiling masses, which make the air

[1] Hemans.

white and thick as with snow, only the flakes
are a foot or two long each : the surges themselves are full of foam in their very bodies
underneath, making them white all through,
as the water is under a great cataract; and
their masses, being thus half water and half
air, are torn to pieces by the wind whenever
they rise, and carried away in roaring smoke,
which chokes and strangles like actual water.
Add to this, that when the air has been ex-
hausted of its moisture by long rain, the spray
of the sea is caught by it as described above,
and covers its surface not merely with the
smoke of finely divided water, but with boil-
ing mist; imagine also the low rain-clouds
brought down to the very level of the sea, as
I have often seen them, whirling and flying in
rags and fragments from wave to wave; and
finally, conceive the surges themselves in their
utmost pitch of power, velocity, vastness, and
madness, lifting themselves in precipices and
peaks, furrowed with their whirl of ascent,
through all this chaos, and you will understand that there is indeed no distinction left
between the sea and air; that no object, nor

horizon, nor any landmark or natural evidence of position is left; and the heaven is all spray, and the ocean all cloud, and that you can see no further in any direction than you see through a cataract."

SEA LIFE

The Sea teems with life. The Great Sea Serpent is, indeed, as much a myth as the Kraken of Pontoppidan, but other monsters, scarcely less marvellous, are actual realities. The Giant Cuttle Fish of Newfoundland, though the body is comparatively small, may measure 60 feet from the tip of one arm to that of another. The Whalebone Whale reaches a length of over 70 feet, but is timid and inoffensive. The Cachalot or Sperm Whale, which almost alone among animals roams over the whole ocean, is as large, and much more formidable. It is armed with powerful teeth, and is said to feed mainly on Cuttle Fish, but sometimes on true fishes, or even Seals. When wounded it often attacks boats, and its companions do not hesitate to come to the rescue. In one case, indeed, an American ship was actually attacked, stove in, and sunk by a gigantic male Cachalot.

The Great Roqual is still more formidable, and has been said to attain a length of 120 feet, but this is probably an exaggeration. So far as we know, the largest species of all is Simmond's Whale, which reaches a maximum of 85 to 90 feet.

In former times Whales were frequent on our coasts, so that, as Bishop Pontoppidan said, the sea sometimes appeared as if covered with smoking chimneys, but they have been gradually driven further and further north, and are still becoming rarer. As they retreated man followed, and to them we owe much of our progress in geography. Is it not, however, worth considering whether they might not also be allowed a " truce of God," whether some part of the ocean might not be allotted to them where they might be allowed to breed in peace ? As a mere mercantile arrangement the maritime nations would probably find this very remunerative. The reckless slaughter of Whales, Sea Elephants, Seals, and other marine animals is a sad blot, not only on the character, but on the common sense, of man.

The monsters of the ocean require large quantities of food, but they are supplied abundantly. Scoresby mentions cases in which the sea was for miles tinged of an olive green by a species of Medusa. He calculates that in a cubic mile there must have been 23,888,000,000,000,000, and though no doubt the living mass did not reach to any great depth, still, as he sailed through water thus discoloured for many miles, the number must have been almost incalculable.

This is, moreover, no rare or exceptional case. Navigators often sail for leagues through shoals of creatures, which alter the whole colour of the sea, and actually change it, as Reclus says, into "une masse animée."

Still, though the whole ocean teems with life, both animals and plants are most abundant near the coast. Air-breathing animals, whether mammals or insects, are naturally not well adapted to live far from dry land. Even Seals, though some of them make re-

markable migrations, remain habitually near
the shore. Whales alone are specially modified
so as to make the wide ocean their home. Of
birds the greatest wanderer is the Albatross,
which has such powers of flight that it is said
even to sleep on the wing.

Many Pelagic animals—Jelly-fishes, Molluscs, Cuttle-fishes, Worms, Crustacea, and
some true fishes — are remarkable for having
become perfectly transparent ; their shells,
muscles, and even their blood have lost all
colour, or even undergone the further modification of having become blue, often with
beautiful opalescent reflections. This obviously renders them less visible, and less liable
to danger.

The sea-shore, wherever a firm hold can be
obtained, is covered with Sea-weeds, which
fall roughly into two main divisions, olive-
green and red. the latter colour having a special
relation to light. These Sea-weeds afford
food and shelter to innumerable animals.

The clear rocky pools left by the retiring
tide are richly clothed with green sea-weeds,
while against the sides are tufts of beautiful

filmy red algæ, interspersed with Sea-anemones, — white, creamy, pink, yellow, purple,
with a coronet of blue beads, and of many
mixed colours ; Sponges, Corallines, Starfish,
Limpets, Barnacles, and other shell-fish ;
feathery Zoophytes and Annelides expand their
pink or white disks, while here and there a
Crab scuttles across ; little Fish or Shrimps
timidly come out from crevices in the rocks,
or from among the fronds of the sea-weeds, or
hastily dart from shelter to shelter; each
little pool is, in fact, a miniature ocean in
itself, and the longer one looks the more and
more one will see.

The dark green and brown sea-weeds do
not live beyond a few — say about 15 —
fathoms in depth. Below them occur delicate
scarlet species, with Corallines and a different
set of shells, Sea-urchins, etc. Down to about
100 fathoms the animals and plants are still
numerous and varied. But they gradually
diminish in numbers, and are replaced by new
forms.

To appreciate fully the extreme loveliness
of marine animals they must be seen alive.

"A tuft of Sertularia, laden with white, or brilliantly tinted Polypites," says Hincks, "like blossoms on some tropical tree, is a perfect marvel of beauty. The unfolding of a mass of Plumularia, taken from amongst the miscellaneous contents of the dredge, and thrown into a bottle of clear sea-water, is a sight which, once seen, no dredger will forget. A tree of Campanularia, when each one of its thousand transparent calycles — itself a study of form — is crowned by a circlet of beaded arms, drooping over its margin like the petals of a flower, offers a rare combination of the elements of beauty.

The rocky wall of some deep tidal pool, thickly studded with the long and slender stems of Tubularia, surmounted by the bright rose-coloured heads, is like the gay parterre of a garden. Equally beautiful is the dense growth of Campanularia, covering (as I have seen it in Plymouth Sound) large tracts of the rock, its delicate shoots swaying to and fro with each movement of the water, like trees in a storm, or the colony of Obelia on the waving frond of the tangle looking almost ethereal in its grace, transparency, and delicacy, as seen against the coarse dark surface that supports it."

Few things are more beautiful than to look down from a boat into transparent water. At the bottom wave graceful sea-weeds, brown, green, or rose-coloured, and of most varied forms; on them and on the sands or rocks rest starfishes, mollusca, crustaceans, Sea-anemones, and innumerable other animals of strange forms and varied colours; in the clear water float or dart about endless creatures; true fishes, many of them brilliantly coloured; Cuttle-fishes like bad dreams; Lobsters and Crabs with graceful, transparent Shrimps; Worms swimming about like living ribbons, some with thousands of coloured eyes, and Medusæ like living glass of the richest and softest hues, or glittering in the sunshine with all the colours of the rainbow.

And on calm, cool nights how often have I stood on the deck of a ship watching with wonder and awe the stars overhead, and the sea-fire below, especially in the foaming, silvery wake of the vessel, where often suddenly appear globes of soft and lambent light, given out perhaps from the surface of some large Medusa.

" A beautiful white cloud of foam," says Coleridge, "at momently intervals coursed by the side of the vessel with a roar, and little stars of flame danced and sparkled and went out in it; and every now and then light detachments of this white cloud-like foam darted off from the vessel's side, each with its own small constellation, over the sea, and scoured out of sight like a Tartar troop over a wilderness."

Fish also are sometimes luminous. The Sun-fish has been seen to glow like a white-hot cannon-ball, and in one species of Shark (Squalus fulgens) the whole surface sometimes gives out a greenish lurid light which makes it a most ghastly object, like some great ravenous spectre.

THE OCEAN DEPTHS

The Land bears a rich harvest of life, but only at the surface. The Ocean, on the contrary, though more richly peopled in its upper layers, which swarm with such innumerable multitudes of living creatures that they are, so to say, almost themselves alive — teems throughout with living beings.

The deepest abysses have a fauna of their own, which makes up for the comparative scantiness of its numbers, by the peculiarity and interest of their forms and organisation. The middle waters are the home of various Fishes, Medusæ, and animalcules, while the upper layers swarm with an inexhaustible variety of living creatures.

It used to be supposed that the depths of the Ocean were destitute of animal life, but recent researches, and especially those made during our great national expedition in the "Challenger," have shown that this is not the case, but that the Ocean depths have a wonderful and peculiar life of their own. Fish have been dredged up even from a depth of 2750 fathoms.

The conditions of life in the Ocean depths are very peculiar. The light of the sun cannot penetrate beyond about two hundred fathoms; deeper than this complete darkness prevails. Hence in many species the eyes have more or less completely disappeared.

Sir Wyville Thomson mentions a kind of Crab (Ethusa granulata), which when living near the surface has well developed eyes; in deeper water, 100 to 400 fathoms, eyestalks are present, but the animal is apparently blind, the eyes themselves being absent; while in specimens from a depth of 500-700 fathoms the eyestalks themselves have lost their special character, and have become fixed, their terminations being combined into a strong, pointed beak.

In other deep sea creatures, on the contrary, the eyes gradually become more and more developed, so that while in some species the eyes gradually dwindle, in others they become unusually large.

Many of the latter species may be said to be a light to themselves, being provided with a larger or smaller number of curious luminous organs. The deep sea fish are either silvery, pink, or in many cases black, sometimes relieved with scarlet, and when the luminous organs flash out must present a very remarkable appearance.

We have still much to learn as to the structure and functions of these organs, but there are cases in which their use can be surmised with some probability. The light is evidently under the will of the fish.[1] It is easy to imagine a Photichthys (Light Fish) swimming in the black depths of the Ocean, suddenly flashing out light from its luminous organs, and thus bringing into view any prey which may be near; while, if danger is disclosed, the light is again at once extinguished. It may be observed that the largest of these organs is in this species situated just under the eye, so that the fish is actually provided with a bull's eye lantern. In other cases the light may rather serve as a defence, some having, as, for instance, in the genus Scopelus, a pair of large ones in the tail, so that "a strong ray of light shot forth from the stern-chaser may dazzle and frighten an enemy."

In other cases they appear to serve as

[1] Gunther, *History of Fishes.*

lures. The "Sea-devil" or "Angler" of
our coasts has on its head three long, very
flexible, reddish filaments, while all round its
head are fringed appendages, closely resembling fronds of sea-weed. The fish conceals
itself at the bottom, in the sand or among
sea-weed, and dangles the long filaments in
front of its mouth. Little fishes, taking these
filaments for worms, unsuspectingly approach,
and thus fall victims.

Several species of the same family live at
great depths, and have very similar habits.
A mere red filament would be invisible in the
dark and therefore useless. They have, however, developed a luminous organ, a living
"glow-lamp," at the end of the filament,
which doubtless proves a very effective lure.

In the great depths, however, fish are comparatively rare. Nor are Molluscs much more
abundant. Sea-urchins, Sea Slugs, and Starfish are more numerous, and on one occasion
20,000 specimens of an Echinus were brought
up at a single haul. True corals are rare, nor
are Hydrozoa frequent. though a gaint species,
allied to the little Hydra of our ponds but

upwards of 6 feet in height, has more than
once been met with. Sponges are numerous,
and often very beautiful. The now well
known Euplectella, " Venus's Flower-basket,"
resembles an exquisitely delicate fabric woven
in spun silk ; it is in the form of a gracefully
curved tube, expanding slightly upwards and
ending in an elegant frill. The wall is formed
of parallel bands of glassy siliceous fibres,
crossed by others at right angles, so as to
form a square meshed net. These sponges
are anchored on the fine ooze by wisps of
glassy filaments, which often attain a considerable length. Many of these beautiful
organisms. moreover, glow when alive with
a soft diffused light, flickering and sparkling
at every touch. What would one not give
to be able to wander a while in these wonderful regions !

It is curious that no plants, so far as we
know, grow in the depths of the Ocean, or,
indeed, as far as our present information goes,
at a greater depth than about 100 fathoms.

As regards the nature of the bottom itself.
it is in the neighbourhood of land mainly

composed of materials, brought down by
rivers or washed from the shore, coarser near
the coast, and tending to become finer and
finer as the distance increases and the water
deepens. The bed of the Atlantic from 400
to 2000 fathoms is covered with an ooze, or
very fine chalky deposit, consisting to a great
extent of minute and more or less broken
shells, especially those of Globigerina. At
still greater depths the carbonate of lime
gradually disappears, and the bottom consists
of fine red clay, with numerous minute particles, some of volcanic, some of meteoric, origin,
fragments of shooting stars, over 100,000,000
of which are said to strike the surface of our
earth every year. How slow the process of
deposition must be, may be inferred from the
fact that the trawl sometimes brings up many
teeth of Sharks and ear-bones of Whales (in
one case no less than 600 teeth and 100 ear-
bones), often semi-fossil, and which from their
great density had remained intact for ages,
long after all the softer parts had perished
and disappeared.

The greatest depth of the Ocean appears

to coincide roughly with the greatest height
of the mountains.   There are indeed cases
recorded in which it is said that "no bottom"
was found even at 39,000 feet.   It is, however, by no means easy to sound at such great
depths, and it is now generally considered
that these earlier observations are untrust-
worthy.   The greatest depth known in the
Atlantic is 3875 fathoms — a little to the
north of the Virgin Islands, but the soundings as yet made in the deeper parts of the
Ocean are few in number, and it is not to be
supposed that the greatest depth has yet been
ascertained.

CORAL ISLANDS

In many parts of the world the geography
itself has been modified by the enormous development of animal life.   Most islands fall
into one of three principal categories:

Firstly, Those which are in reality a part
of the continent near which they lie, being
connected by comparatively shallow water,
and standing to the continent somewhat in

the relation of planets to the sun; as, for
instance, the Cape de Verde Islands to Africa,
Ceylon to India, or Tasmania to Australia.

Secondly, Volcanic islands; and

Thirdly, Those which owe their origin to
the growth of Coral reefs.

Fig. 49. — Whitsunday Island.

Coral islands are especially numerous in
the Indian and Pacific Oceans, where there
are innumerable islets, in the form of rings,
or which together form rings, the rings themselves being sometimes made up of ringlets.
These "atolls" contain a circular basin of
yellowish green, clear, shallow water, while
outside is the dark blue deep water of the
Ocean. The islands themselves are quite low,
with a beach of white sand rising but a few

feet above the level of the water, and bear generally groups of tufted Cocoa Palms.

It used to be supposed that these were the summits of submarine volcanoes on which the coral had grown. But as the reef-making coral does not live at greater depths than about twenty-five fathoms, the immense number of these reefs formed an almost insuperable objection to this theory. The Laccadives and Maldives for instance — meaning literally the "lac of or 100,000 islands," and the "thousand islands" — are a series of such atolls, and it was impossible to imagine so great a number of craters, all so nearly of the same altitude.

In shallow tracts of sea, coral reefs no doubt tend to assume the well-known circular form, but the difficulty was to account for the numerous atolls which rise to the surface form the abysses of the ocean, while the coral-forming zoophytes can only live near the surface.

Darwin showed that so far from the ring of corals resting on a corresponding ridge of rocks, the lagoons, on the contrary, now occupy the place which was once the highest land. He pointed out that some lagoons, as for instance that of Vanikoro, contain an island in the middle; while other islands, such as Tahiti, are surrounded by a margin of smooth water separated from the ocean by a coral reef. Now if we suppose that Tahiti were to sink slowly it would gradually approximate to the condition of Vanikoro; and if Vanikoro gradually sank, the central island would disappear, while on the contrary the growth of the coral might neutralise the subsidence of the reef, so that we should have simply an atoll with its lagoon. The same considerations explain the origin of the "barrier reefs," such as that which runs for nearly a thousand miles, along the north-east coast of Australia. Thus Darwin's theory explains the form and the approximate identity of altitude of these coral islands. But it does more than this, because it shows that there are great areas in process of subsidence, which though slow, is of great importance in physical geography.

The lagoon islands have received much attention ; which " is not surprising, for every one must be struck with astonishment, when he first beholds one of these vast rings of coral-rock, often many leagues in diameter, here and there surmounted by a low verdant island with dazzling white shores, bathed on the outside by the foaming breakers of the ocean, and on the inside surrounding a calm expanse of water, which, from reflection is generally of a bright but pale green colour. The naturalist will feel this astonishment more deeply after having examined the soft and almost gelatinous bodies of these apparently insignificant coral-polypifers, and when he knows that the solid reef increases only on the outer edge, which day and night is lashed by the breakers of an ocean never at rest. Well did François Pyrard de Laval, in the year 1605 exclaim. ' C'est une merveille de voir chacun de ces atollons, environné d'un grand banc de pierre tout autour, n'y ayant point d'artifice humain.' " [1]

Of the enchanting beauty of the coral beds

[1] Darwin, *Coral Reefs.*

themselves we are assured that language conveys no adequate idea. "There were corals," says Prof. Ball, "which, in their living state, are of many shades of fawn, buff, pink, and blue, while some were tipped with a magenta-like bloom. Sponges which looked as hard as stone spread over wide areas, while sprays of coralline added their graceful forms to the picture. Through the vistas so formed, golden-banded and metallic-blue fish meandered, while on the patches of sand here and there Holothurias and various mollusca and crustaceans might be seen slowly crawling."

Abercromby also gives a very graphic description of a Coral reef. "As we approached," he says, "the roaring surf on the outside, fingery lumps of beautiful live coral began to appear of the palest lavender-blue colour; and when at last we were almost within the spray, the whole floor was one mass of living branches of coral.

"But it is only when venturing as far as is prudent into the water, over the outward edge of the great sea wall, that the true character of the reef and all the beauties of the ocean can be really seen. After walking over a flat uninteresting tract of nearly bare rock, you look down and see a steep irregular wall, expanding deeper into the ocean than the eye can follow, and broken into lovely grottoes and holes and canals, through which small resplendent fish of the brightest blue or gold flit fitfully between the lumps of coral. The sides of these natural grottoes are entirely covered with endless forms of tender-coloured coral, but all beautiful, and all more or less of the fingery or branching species, known as madrepores. It is really impossible to draw or describe the sight, which must be taken with all its surroundings as adjuncts." [1]

The vegetation of these fairy lands is also very lovely; the Coral tree (Erythrina) with light green leaves and bunches of scarlet blossoms, the Cocoa-nut always beautiful, the breadfruit, the graceful tree ferns, the Barringtonia, with large pink and white flowers, several species of Convolvulus, and many others unknown to us even by name.

[1] Abercromby, *Seas and Skies in many Latitudes.*

THE SOUTHERN SKIES

In considering these exquisite scenes, the beauty of the Southern skies must not be omitted. "From the time we entered the torrid zone," says Humboldt, "we were never wearied with admiring, every night, the beauty of the southern sky, which, as we advanced towards the south, opened new constellations to our view. We feel an indescribable sensation, when, on approaching the equator, and particularly on passing from one hemisphere to the other, we see those stars which we have contemplated from our infancy, progressively sink, and finally disappear. Nothing awakens in the traveller a livelier remembrance of the immense distance by which he is separated from his country, than the aspect of an unknown firmament. The grouping of the stars of the first magnitude, some scattered nebulæ rivalling in splendour the milky way, and tracts of space remarkable for their extreme blackness, give a particular physiognomy to the southern sky. This sight fills with admiration even those, who, uninstructed in the branches of accurate science, feel the same emotions of delight in the contemplation of the heavenly vault, as in the view of a beautiful landscape, or a majestic river. A traveller has no need of being a botanist to recognise the torrid zone on the mere aspect of its vegetation; and, without having acquired any notion of astronomy, he feels he is not in Europe, when he sees the immense constellation of the Ship, or the phosphorescent clouds of Magellan, arise on the horizon. The heaven and the earth, in the equinoctial regions, assume an exotic character."

"The sunsets in the Eastern Archipelago," says H. O. Forbes,[1] "were scenes to be remembered for a life-time. The tall cones of Sibissie and Krakatoa rose dark purple out of an unruffled golden sea, which stretched away to the south-west, where the sun went down; over the horizon gray fleecy clouds lay in banks and streaks, above them pale blue lanes of sky, alternating with orange bands, which higher up gave place to an expanse of

[1] *A Naturalist's Wanderings in the Eastern Archipelago.*

red stretching round the whole heavens.
Gradually as the sun retreated deeper and
deeper, the sky became a marvellous golden
curtain, in front of which the gray clouds
coiled themselves into weird forms before
dissolving into space. . . ."

THE POLES

The Arctic and Antarctic regions have
always exercised a peculiar fascination over
the human mind. Until now every attempt
to reach the North Pole has failed, and the
South has proved even more inaccessible.
In the north, Parry all but reached lat. 83;
in the south no one has penetrated beyond
lat. 71.11. And yet, while no one can say
what there may be round the North Pole, and
some still imagine that open water might be
found there, we can picture to ourselves the
extreme South with somewhat more confidence.

Whenever ships have sailed southwards,
except at a few places where land has been
met with, they have come at last to a wall of
ice, from fifty to four hundred feet high. In

those regions it snows, if not incessantly, at
least very frequently, and the snow melts but
little.   As far as the eye can reach nothing is
to be seen but snow.   Now this snow must
gradually accumulate, and solidify into ice,
until it attains such a slope that it will move
forward as a glacier.   The enormous Icebergs
of the Southern Ocean, moreover, show that
it does so, and that the snow of the extreme
south, after condensing into ice, moves slowly
outward and at length forms a wall of ice,
from which Icebergs, from time to time,
break away.   We do not exactly know what,
under such circumstances, the slope would
be; but Mr. Croll points out that if we take
it at only half a degree, and this seems quite
a minimum, the Ice cap at the South Pole
must be no less than twelve miles in thickness.
It is indeed probably even more, for some of
the Southern tabular icebergs attain a height
of eight hundred, or even a thousand feet
above water, indicating a total thickness of
the ice sheet even at the edge, of over a mile.

Sir James Ross mentions that — "Whilst
measuring some angles for the survey near

Mount Lubbock an island suddenly appeared, which he was quite sure was not to be seen two or three hours previously. He was much astonished, but it eventually turned out to be a large iceberg, which had turned over, and so exposed a new surface covered with earth and stones."

The condition of the Arctic regions is quite different. There is much more land, and no such enormous solid cap of ice. Spitzbergen, the land of " pointed mountains," is said to be very beautiful. Lord Dufferin describes his first view of it as " a forest of thin lilac peaks, so faint, so pale, that had it not been for the gem-like distinctness of their outline one could have deemed them as unsubstantial as the spires of Fairy-land."

It is, however, very desolate; scarcely any vegetation excepting a dark moss, and even this goes but a little way up the mountain side. Scoresby ascended one of the hills near Horn Sound, and describes the view as " most extensive and grand. A fine sheltered bay was seen to the east of us, an arm of the same on the north-east, and the sea, whose glassy surface was unruffled by a breeze, formed an immense expanse on the west; the glaciers, rearing their proud crests almost to the tops of mountains between which they were lodged, and defying the power of the solar beams, were scattered in various directions about the sea-coast and in the adjoining bays. Beds of snow and ice filling extensive hollows, and giving an enamelled coat to adjoining valleys, one of which, commencing at the foot of the mountain where we stood, extended in a continual line towards the north, as far as the eye could reach — mountain rising above mountain, until by distance they dwindled into insignificance, the whole contrasted by a cloudless canopy of deepest azure, and enlightened by the rays of a blazing sun, and the effect, aided by a feeling of danger, seated as we were on the pinnacle of a rock almost surrounded by tremendous precipices — all united to constitute a picture singularly sublime."

One of the glaciers of Spitzbergen is 11 miles in breadth when it reaches the sea-coast, the highest part of the precipitous front adjoining the sea being over 400 feet, and it extends far upwards towards the summit of the mountain. The surface forms an inclined plane of smooth unsullied snow, the beauty and brightness of which render it a conspicuous landmark on that inhospitable shore. From the perpendicular face great masses of ice from time to time break away,

> Whose blocks of sapphire seem to mortal eye
> Hewn from cærulean quarries of the sky.[1]

Field ice is comparatively flat, though it may be piled up perhaps as much as 50 feet. It is from glaciers that true icebergs, the beauty and brilliance of which Arctic travellers are never tired of describing, take their origin.

The attempts to reach the North Pole have cost many valuable lives; Willoughby and Hudson, Behring and Franklin, and many other brave mariners; but yet there are few expeditions more popular than those to " the Arctic," and we cannot but hope that it is still reserved for the British Navy after so many gallant attempts at length to reach the North Pole.

[1] Montgomery.

CHAPTER X

THE STARRY HEAVENS

A man can hardly lift up his eyes towards the heavens without wonder and veneration, to see so many millions of radiant lights, and to observe their courses and revolutions, even without any respect to the common good of the Universe. — SENECA.

CHAPTER X

THE STARRY HEAVENS

MANY years ago I paid a visit to Naples, and ascended Vesuvius to see the sun rise from the top of the mountain. We went up to the Observatory in the evening and spent the night outside. The sky was clear; at our feet was the sea, and round the bay the lights of Naples formed a lovely semicircle. Far more beautiful, however, were the moon and the stars overhead; the moon throwing a silver path over the water, and the stars shining in that clear atmosphere with a brilliance which I shall never forget.

For ages and ages past men have admired the same glorious spectacle, and yet neither the imagination of Man nor the genius of Poetry had risen to the truer and grander

conceptions of the Heavens for which we
are indebted to astronomical Science. The
mechanical contrivances by which it was
attempted to explain the movements of the
heavenly bodies were clumsy and prosaic
when compared with the great discovery of
Newton. Ruskin is unjust I think when he
says "Science teaches us that the clouds are
a sleety mist; Art, that they are a golden
throne." I should be the last to disparage
the debt we owe to Art, but for our knowledge, and even more, for our appreciation,
feeble as even yet it is, of the overwhelming
grandeur of the Heavens, we are mainly indebted to Science.

There is scarcely a form which the fancy of
Man has not sometimes detected in the clouds,
—chains of mountains, splendid cities, storms
at sea, flights of birds, groups of animals,
monsters of all kinds,—and our superstitious
ancestors often terrified themselves by fantastic visions of arms and warriors and battles
which they regarded as portents of coming
calamities. There is hardly a day on which
Clouds do not delight and surprise us by their

THE MOON. *To face page* 377.

forms and colours. They belong, however, to our Earth, and I must now pass on to the heavenly bodies.

THE MOON

The Moon is the nearest, and being the nearest, appears to us, with the single exception of the Sun, the largest, although it is in reality one of the smallest, of the heavenly bodies. Just as the Earth goes round the Sun, and the period of revolution constitutes a year, so the Moon goes round the Earth approximately in a period of one month. But while we turn on our axis every twenty-four hours, thus causing the alternation of light and darkness — day and night — the Moon takes a month to revolve on hers, so that she always presents the same, or very nearly the same, surface to us.

Seeing her as we do, not like the Sun and Stars, by light of her own, but by the reflected light of the Sun, her form appears to change, because the side upon which the Sun shines is not always that which we see. Hence the "phases" of the Moon, which add so much to her beauty and interest.

Who is there who has not watched them with admiration? " We first see her as an exquisite crescent of pale light in the western sky after sunset. Night after night she moves further and further to the east, until she becomes full, and rises about the same time that the Sun sets. From the time of full moon the disc of light begins to diminish, until the last quarter is reached. Then it is that the Moon is seen high in the heavens in the morning. As the days pass by, the crescent shape is again assumed. The crescent wanes thinner and thinner as the Moon draws closer to the Sun. Finally, she becomes lost in the overpowering light of the Sun, again to emerge as the new moon, and again to go through the same cycle of changes." [1]

But although she is so small the Moon is not only, next to the Sun, by far the most beautiful, but also for us the most important, of the heavenly bodies. Her attraction, aided by that of the Sun, causes the tides, which

[1] Ball, *Story of the Heavens.*

are of such essential service to navigation. They carry our vessels in and out of port, and, indeed, but for them many of our ports would themselves cease to exist, being silted up by the rivers running into them. The Moon is also of invaluable service to sailors by enabling them to determine where they are, and guiding them over the pathless waters.

The geography of the Moon, so far as concerns the side turned towards us, has been carefully mapped and studied, and may almost be said to be as well known as that of our own earth. The scenery is in a high degree weird and rugged ; it is a great wilderness of extinct volcanoes, and, seen with even a very moderate telescope, is a most beautiful object. The mountains are of great size. Our loftiest mountain, Mount Everest, is generally stated as about 29,000 feet in height. The mountains of the Moon reach an altitude of over 42,000, but this reckons to the lowest depression, and it must be remembered that we reckon the height of mountains to the sea level only. Several of the craters on the Moon have a diameter of 40 or 50 — one of

them even as much as 78 — miles. Many
also have central cones, closely resembling
those in our own volcanic regions. In some
cases the craters are filled nearly to the brim
with lava. The volcanoes seem, however, to
be all extinct; and there is not a single case

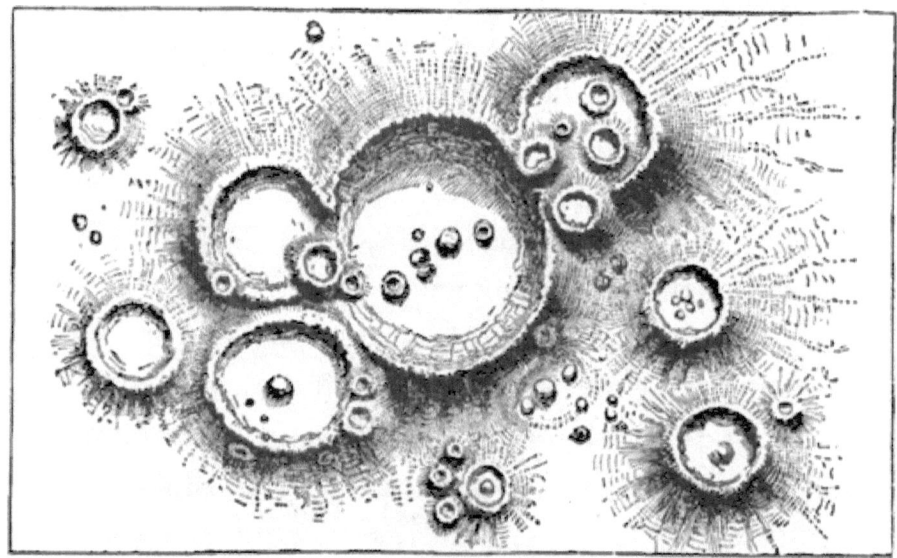

Fig. 50. — A group of Lunar Volcanoes.

in which we have conclusive evidence of any
change in a lunar mountain.

The Moon, being so much smaller than the
earth, cooled, of course, much more rapidly,
and it is probable that these mountains are
millions of years old — much older than many

of our mountain chains.   Yet no one can look
at a map of the Moon without being struck
with the very rugged character of its mountain scenery.   This   is mainly due to the
absence of air and water.   To these two
mighty agencies, not merely " the cloud-
capped towers, the gorgeous palaces, the
solemn temples," but the very mountains
themselves, are inevitable victims.   Not
merely storms and hurricanes, but every
gentle shower, every fall of snow, tends to
soften our scenery and lower the mountain
peaks.   These agencies are absent from the
Moon, and the mountains stand to-day just
as they were formed millions of years ago.

But though we find on our own globe (see,
for instance, Fig. 21) volcanic regions closely
resembling those of the Moon, there are other
phenomena on the Moon's surface for which
our earth presents as yet no explanation.
From Tycho, for instance, a crater 17,000
feet high and 50 miles across, a number of
rays or streaks diverge, which for hundreds,
or in some cases two or three thousand, miles
pass straight across plains, craters, and moun-

tains. The true nature of these streaks is not yet understood.

THE SUN

The Sun is more than 400 times as distant as the Moon; a mighty glowing globe, infinitely hotter than any earthly fiery furnace, 300,000 times as heavy, and 1,000,000 times as large as the earth. Its diameter is 865,000 miles, and it revolves on its axis in between 25 and 26 days. Its distance is 92,500,000 miles. And yet it is only a star, and by no means one of the first magnitude.

The surface of the Sun is the seat of violent storms and tempests. From it gigantic flames, consisting mainly of hydrogen, flicker and leap. Professor Young describes one as being, when first observed, 40,000 miles high. Suddenly it became very brilliant, and in half an hour sprang up 40,000 more. For another hour it soared higher and higher, reaching finally an elevation of no less than 350,000 miles, after which it slowly faded away, and in a couple of hours had entirely disappeared. This was no doubt an exceptional case, but a height of 100,000 miles is not unusual, and the velocity frequently reaches 100 miles in a second.

The proverbial spots on the Sun in many respects resemble the appearances which would be presented if a comparatively dark central mass was here and there exposed by apertures through the more brilliant outer gases, but their true nature is still a matter of discussion.

During total eclipses it is seen that the Sun is surrounded by a "corona," or aureola of light, consisting of radiant filaments, beams, and sheets of light, which radiate in all directions, and the true nature of which is still doubtful.

Another stupendous problem connected with the Sun is the fact that, as geology teaches us, it has given off nearly the same quantity of light and heat for millions of years. How has this come to pass? Certainly not by any process of burning such as we are familiar with. Indeed, if the heat of the Sun were due to combustion it would be burnt up in 6000 years. It has been suggested that the meteors, which fall in showers on to the Sun, replace the heat which is emitted. To some slight extent perhaps they do so, but the main cause seems to be the slow condensation of the Sun itself. Mathematicians tell us that a contraction of about 220 feet a year would account for the whole heat emitted, and as the present diameter of the Sun is about 860,000 miles, the potential store of heat is still enormous.

To the Sun we owe our light and heat; it is not only the centre of our planetary system, it is the source and ruler of our lives. It draws up water from the ocean, and pours it down in rain to fill the rivers and refresh the plants; it raises the winds, which purify the air and waft our ships over the seas; it draws our carriages and drives our steam-engines, for coal is but the heat of former ages stored up for our use; animals live and move by the Sun's warmth; it inspires the song of birds, paints the flowers, and ripens the fruit. Through it the trees grow. For the beauties of nature, for our food and drink, for our clothing, for our light and life, for the very possibility of our existence, we are indebted to the Sun.

What is the Sun made of ? Comte mentioned as a problem, which it was impossible that man could ever solve, any attempt to determine the chemical composition of the heavenly bodies. " Nous concevons," he said, " la possibilité de déterminer leurs formes, leurs distances, leurs grandeurs, et leurs mouvements, tandis que nous ne saurions jamais étudier par aucun moyen leur composition chimique ou leur structure minéralogique." To do so might well have seemed hopeless, and yet the possibility has been proved, and a beginning has been made. In the early part of this century Wollaston observed that the bright band of colours thrown by a prism, and known as the spectrum, was traversed by dark lines, which were also discovered, and described more in detail, by Fraunhofer, after whom they are generally called " Fraunhofer's lines." The next step was made by Wheatstone, who showed that the spectrum formed by incandescent vapours was formed of bright lines, which differed for each substance, and might, therefore, be used as a convenient mode of analysis. In fact, by this process several new substances have actually been discovered. These bright lines were found on comparison to coincide with the dark lines in the spectrum, and to Kirchhoff and Bunsen is due the credit of applying this method of research to astronomical science. They arranged their apparatus so that one-half was lighted by the Sun, the other by the incandescent gas they were examining. When the vapour of sodium was treated in this way they found that the bright line in the flame of soda exactly coincided with a line in the Sun's spectrum. The conclusion was obvious; there is sodium in the Sun. It must, indeed, have been a glorious moment when the thought flashed upon them; and the discovery, with its results, is one of the greatest triumphs of human genius.

The Sun has thus been proved to contain hydrogen, sodium, barium, magnesium, calcium, aluminium, chromium, iron, nickle, manganese, titanium, cobalt, lead, zinc, copper, cadmium, strontium, cerium, uranium, potassium, etc., in all 36 of our terrestrial elements, while as regards some others the evidence is not conclusive. We cannot as yet say that any of our elements are absent, nor though there are various lines which cannot as yet be certainly referred to any known substance, have we clear proof that the Sun contains any element which does not exist on our earth. On the whole, then, the chemical composition of the Sun appears closely to resemble that of our earth.

THE PLANETS

The Syrian shepherds watching their flocks by night long ago noticed — and they were probably not the first — that there were five stars which did not follow the regular course of the rest, but, apparently at least, moved about irregularly. These they appropriately named Planets, or wanderers.

Further observations have shown that this irregularity of their path is only apparent, and that, like our own Earth, they really revolve round the Sun. To the five first observed —

Mercury, Venus, Mars, Jupiter, and Saturn — two large ones, Uranus and Neptune, and a group of minor bodies, have since been added.

The following two diagrams give the relative orbits of the Planets.

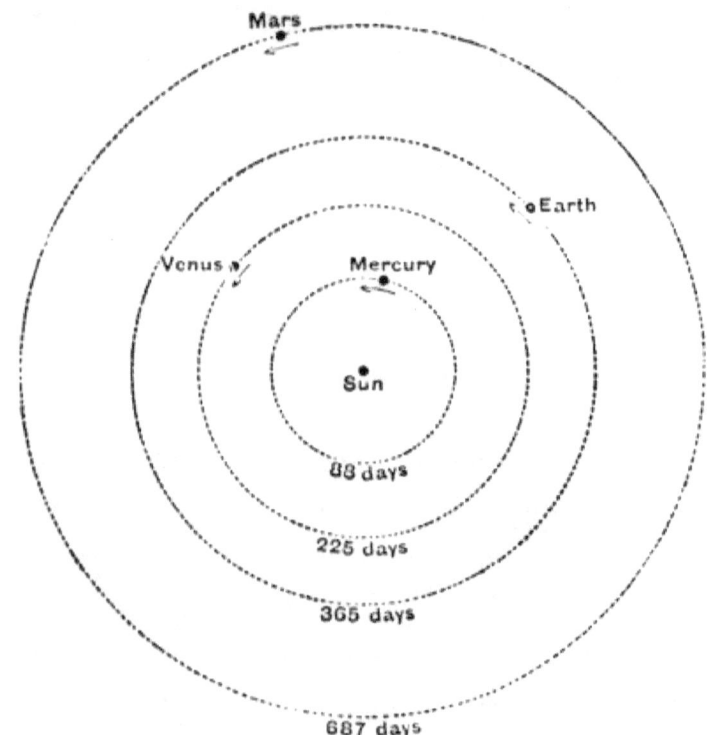

Fig. 51. — Orbits of the inner Planets.

MERCURY

It is possible, perhaps probable, that there may be an inner Planet, but, so far as we

know for certain, Mercury is the one nearest
to the Sun, its average distance being
36,000,000 miles. It is much smaller than
the Earth, its weight being only about $\frac{1}{24}$th

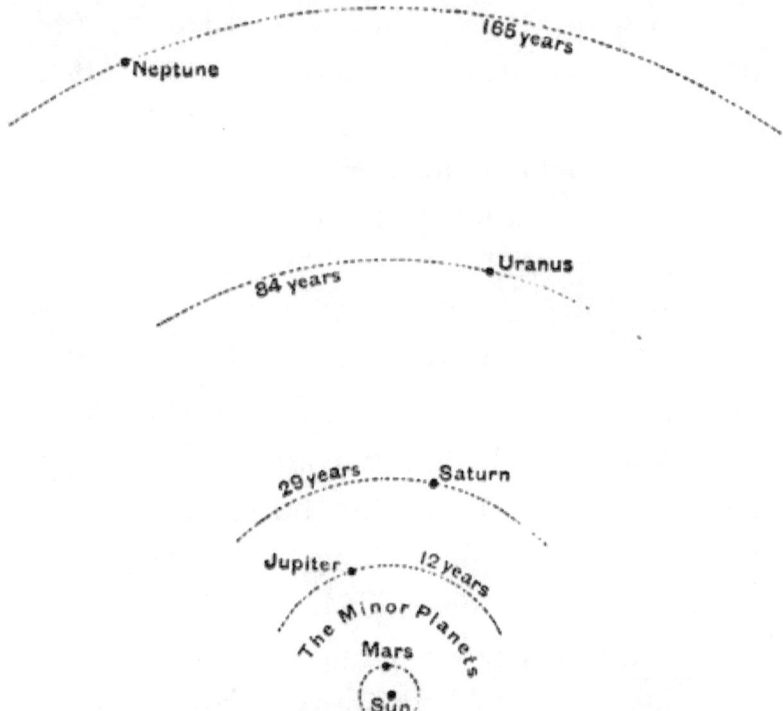

Fig. 52. — Relative distances of the Planets from the Sun.

of ours. Mercury is a shy though beautiful
object, for being so near the Sun it is not
easily visible ; it may, however, generally be
seen at some time or other during the year as
a morning or evening star.

VENUS

The true morning or evening star, however, is Venus — the peerless and capricious Venus.

Venus, perhaps, " has not been noticed, not been thought of, for many months. It is a beautifully clear evening ; the sun has just set. The lover of nature turns to admire the sunset, as every lover of nature will. In the golden glory of the west a beauteous gem is seen to glisten ; it is the evening star, the planet Venus. A week or two later another beautiful sunset is seen, and now the planet is no longer a glistening point low down ; it has risen high above the horizon, and continues a brilliant object long after the shades of night have descended. Again a little longer and Venus has gained its full brilliancy and splendour. All the heavenly host — even Sirius and Jupiter — must pale before the splendid lustre of Venus, the unrivalled queen of the firmament." [1]

Venus is about as large as our Earth, and when at her brightest outshines about fifty

[1] Ball, *Story of the Heavens.*

times the most brilliant star. Yet, like all
the other planets, she glows only with the
reflected light of the Sun, and consequently
passes through phases like those of the Moon,
though we cannot see them with the naked
eye. To Venus also owe we mainly the power
of determining the distance, and consequently
the magnitude, of the Sun.

THE EARTH

Our own Earth has formed the subject of
previous chapters. I will now, therefore, only
call attention to her movements, in which, of
course, though unconsciously, we participate.
In the first place, the Earth revolves on her
axis in 24 hours. Her circumference at the
tropics is 24,000 miles. Hence a person at the
tropics is moving in this respect at the rate of
1000 miles an hour, or over 16 miles a
minute.

But more than this, astronomers have
ascertained that the whole solar system is
engaged in a great voyage through space,
moving towards a point on the constellation

of Hercules at the rate of at least 20,000 miles an hour, or over 300 miles a minute.[1]

But even more again, we revolve annually round the Sun in a mighty orbit 580,000,000 miles in circumference. In this respect we are moving at the rate of no less than 60,000 miles an hour. or 1000 miles a minute — a rate far exceeding of course, in fact by some 100 times, that of a cannon ball.

How few of us know, how little we any of us realise. that we are rushing through space with such enormous velocity.

MARS

To the naked eye Mars appears like a ruddy star of the first magnitude. It has two satellites, which have been happily named Phobos and Deimos — Fear and Dismay. It is little more than half as large as the Earth. and, though generally far more distant, it sometimes approaches us within 35,000,000 miles. This has enabled us to study its physical structure. It seems very probable

[1] Some authorities estimate it even higher.

that there is water in Mars, and the two poles are tipped with white, as if capped by ice and snow.   It presents also a series of remarkable parallel lines, the true nature of which is not yet understood.

THE MINOR PLANETS

A glance at Figs. 51 and 52 will show that the distances of the Planets from the Sun follow a certain rule.

If we take the numbers 0, 3, 6, 12, 24, 48, 96, each one (after the second) the double of that preceding, and add four, we have the series.

4   7   10   16   28   52   100

Now the distances of the Planets from the Sun are as follow : —

Mercury.	Venus.	Earth.	Mars.	Jupiter.	Saturn.
3.9	7.2	10	15.2	52.9	95.4

For this sequence, which was first noticed by Bode, and is known as Bode's law, no explanation can yet be given. It was of course at once observed that between Mars and Jupiter one place is vacant, and it has now been ascertained that this is occupied by a zone of Minor Planets, the first of which was discovered by Piazzi on January 1, 1801, a worthy prelude to the succession of scientific discoveries which form the glory of our century. At present over 300 are known, but certainly these are merely the larger among an immense number, some of them doubtless mere dust.

JUPITER

Beyond the Minor Planets we come to the stupendous Jupiter, containing 300 times the mass, and being 1200 times the size of our Earth — larger indeed than all the other planets put together. It is probably not solid, and from its great size still retains a large portion of the original heat, if we may use such an expression. Jupiter usually shows a number of belts, supposed to be due to clouds floating over the surface, which have a tendency to arrange themselves in belts or bands, owing to the rotation of the planet. Jupiter has four moons or satellites.

SATURN

Next to Jupiter in size, as in position, comes Saturn, which, though far inferior in dimensions, is much superior in beauty. To the naked eye Saturn appears as a brilliant star, but when Galileo first saw it through a telescope it appeared to him to be composed of three bodies in a line, a central globe with a small one on each side. Huyghens in 1655

Fig. 53. — Saturn.

first showed that in reality Saturn was surrounded by a series of rings (see Fig. 53). Of these there are three, the inner one very faint, and the outer one divided into two by a dark line. These rings are really enormous shoals of minute bodies revolving round the planet, and rendering it perhaps the most marvellous and beautiful of all the heavenly bodies.

While we have one Moon, Mars two, and Jupiter four, Saturn has no less than eight satellites.

URANUS

Saturn was long supposed to be the outermost body belonging to the solar system. In 1781, however, on the 13th March, William Herschel was examining the stars in the constellation of the Twins. One struck him because it presented a distinct disc, while the true fixed stars, however brilliant, are, even with the most powerful telescope, mere points of light. At first he thought it might be a comet, but careful observations showed that it was really a new planet. Though thus discovered by Herschel it had often been seen before, but its true nature was unsuspected. It has a diameter of about 31,700 miles.

Four satellites of Uranus have been discovered, and they present the remarkable peculiarity that while all the other planets and their satellites revolve nearly in one plane, the satellites of Uranus are nearly at right angles, indicating the presence of some local and exceptional influence.

NEPTUNE

The study of Uranus soon showed that it followed a path which could not be accounted for by the influence of the Sun and the other then known planets. It was suspected, therefore, that this was due to some other body not yet discovered. To calculate where such a body must be so as to account for these irregularities was a most complex and difficult, and might have seemed almost a hopeless, task. It was, however, solved almost simultaneously and independently by Adams in this country, and Le Verrier in France.

Neptune, so far as we yet know the outmost of our companions, is 35,000 miles in diameter, and its mean distance from the Sun is 2,780,000,000 miles.

ORIGIN OF THE PLANETARY SYSTEM

The theory of the origin of the Planetary System known as the "Nebular Hypothesis," which was first suggested by Kant, and developed by Herschel and Laplace, may be fairly said to have attained a high degree of probability. The space now occupied by the solar system is supposed to have been filled by a rotating spheroid of extreme tenuity and enormous heat, due perhaps to the collision of two originally separate bodies. The heat, however, having by degrees radiated into space, the gas cooled and contracted towards a centre, destined to become the Sun. Through the action of centrifugal force the gaseous matter also flattened itself at the two poles, taking somewhat the form of a disc. For a certain time the tendency to contract, and the centrifugal force, counterbalanced one another, but at length a time came when the latter prevailed and the outer zone detached itself from the rest of the sphere. One after another similar rings were thrown off, and then breaking up, formed the planets and their satellites.

That each planet and satellite did form originally a ring we still have evidence in the wonderful and beautiful rings of Saturn, which, however, in all probability will eventually form spherical satellites like the rest. Thus then our Earth was originally a part of the Sun, to which again it is destined one day to return. M. Plateau has shown experimentally that by rotating a globe of oil in a mixture of water and spirit having the same density this process may be actually repeated in miniature.

This brilliant, and yet simple, hypothesis is consistent with, and explains many other circumstances connected with the position, magnitude, and movements of the Planets and their satellites.

The Planets, for instance, lie more or less in the same plane, they revolve round the Sun and rotate on their own axis in the same direction — a series of coincidences which cannot be accidental, and for which the theory would account. Again the rate of cooling would of course follow the size; a small body cools more rapidly than a large one. The

Moon is cold and rigid; the Earth is solid at
the surface, but intensely hot within; Jupiter
and Saturn, which are immensely larger, still
retain much of their original heat, and have
a much lower density than the Earth; and
astronomers tell us on other grounds that the
Sun itself is still contracting, and that to this
the maintenance of its temperature is due.

Although, therefore, the Nebular Theory
cannot be said to have been absolutely proved,
it has certainly been brought to a high state
of probability, and is, in its main features,
generally accepted by astronomers.

The question has often been asked whether
any of the heavenly bodies are inhabited, and
as yet it is impossible to give any certain
answer.   It seems *à priori* probable that the
millions of suns which we see as stars must
have satellites, and that some at least of them
may be inhabited.   So far as our own system
is concerned the Sun is of course too hot to
serve as a dwelling-place for any beings with
bodies such as ours.   The same may be said
of Mercury, which is at times probably ten
times as hot as our tropics.   The outer planets

appear to be still in a state of vapour.   The
Moon has no air or water.

Mars is in a condition which most nearly
resembles ours.   All, however, that can be
said is that, so far as we can see, the existence of living beings on Mars is not impossible.

COMETS

The Sun, Moon, and Stars, glorious and
wonderful as they are, though regarded with
great interest, and in some cases worshipped
as deities, excited the imagination of our
ancestors less than might have been expected,
and even now attract comparatively little
attention, from the fact that they are always
with us.   Comets, on the other hand, both as
rare and occasional visitors, from their large
size and rapid changes, were regarded in
ancient times with dread and with amazement.

Some Comets revolve round the Sun in
ellipses, but many, if not the majority, are
visitors indeed, for having once passed round

the Sun they pass away again into space, never to return.

The appearance which is generally regarded as characteristic of a Comet is that of a head with a central nucleus and a long tail. Many, however, of the smaller ones possess no tail, and in fact Comets present almost innumerable differences. Moreover the same Comet changes rapidly, so that when they return, they are identified not in any way by their appearance, but by the path they pursue.

Comets may almost be regarded as the ghosts of heavenly bodies. The heads, in some cases, may consist of separate solid fragments, though on this astronomers are by no means agreed, but the tails at any rate are in fact of almost inconceivable tenuity. We know that a cloud a few hundred feet thick is sufficient to hide, not only the stars, but even the Sun himself. A Comet is thousands of miles in thickness, and yet even extremely minute stars can be seen through it with no appreciable diminution of brightness. This extreme tenuity of comets is

moreover shown by their small weight.
Enormous as they are I remember Sir G.
Airy saying that there was probably more
matter in a cricket ball than there is in a
comet. No one, however, now doubts that
the weight must be measured in tons; but
it is so small, in relation to the size, as to
be practically inappreciable. If indeed they
were comparable in mass even to the planets,
we should long ago have perished. The
security of our system is due to the fact that
the planets revolve round the Sun in one
direction, almost in circles, and very nearly
in the same plane. Comets, however, enter
our system in all directions, and at all angles;
they are so numerous that, as Kepler said,
there are probably more Comets in the sky
than there are fishes in the sea, and but for
their extreme tenuity they would long ago
have driven us into the Sun.

When they first come in sight Comets
have generally no tail; it grows as they
approach the Sun, from which it always
points away. It is no mere optical illusion;
but while the Comet as a whole is attracted

by the Sun, the tail, how or why we know
not, is repelled. When once driven off, moreover, the attraction of the Comet is not sufficient to recall it, and hence perhaps so many
Comets have now no tails.

Donati's Comet, the great Comet of 1858,
was first noticed on the 2d June as a faint
nebulous spot. For three months it remained
quite inconspicuous, and even at the end of
August was scarcely visible to the naked eye.
In September it grew rapidly, and by the
middle of October the tail extended no less
than 40 degrees, after which it gradually
disappeared.

Faint as is the light emitted by Comets,
it is yet their own, and spectrum analysis has
detected the presence in them of carbon,
hydrogen, nitrogen, sodium, and probably of
iron.

Comets then remain as wonderful, and
almost as mysterious, as ever, but we need no
longer regard " a comet as a sign of impend-
ing calamity; we may rather look upon it as
an interesting and a beautiful visitor, which
comes to please us and to instruct us, but

never to threaten or to destroy." [1]   We are free, therefore, to admire them in peace, and beautiful, indeed, they are.

"The most wonderful sight I remember," says Hamerton, "as an effect of calm, was the inversion of Donati's Comet, in the year 1858, during the nights when it was sufficiently near the horizon to approach the rugged outline of Graiganunie, and be reflected beneath it in Loch Awe.   In the sky was an enormous aigrette of diamond fire, in the water a second aigrette, scarcely less splendid, with its brilliant point directed upwards, and its broad, shadowy extremity ending indefinitely in the deep.   To be out on the lake alone, in a tiny boat, and let it rest motionless on the glassy water, with that incomparable spectacle before one, was an experience to be remembered through a lifetime.   I have seen many a glorious sight since that now distant year, but nothing to equal it in the association of solemnity with splendour." [2]

[1] Ball.   [2] Hamerton, *Landscape.*

SHOOTING STARS

On almost any bright night, if we watch a short time some star will suddenly seem to drop from its place, and, after a short plunge, to disappear. This appearance is, however, partly illusory. While true stars are immense bodies at an enormous distance, Shooting Stars are very small, perhaps not larger than a paving stone, and are not visible until they come within the limits of our atmosphere, by the friction with which they are set on fire and dissipated. They are much more numerous on some nights than others. From the 9th to the 11th August we pass through one cluster which is known as the Perseids; and on the 13th and 14th November a still greater group called by astronomers the Leonids. The Leonids revolve round the Sun in a period of 33 years, and in an elliptic orbit, one focus of which is about at the same distance from the Sun as we are, the other at about that of Uranus. The shoal of stars is enormous; its diameter cannot be less than 100,000 miles, and its length many hundreds of thousands.

There are, indeed, stragglers scattered over the whole orbit, with some of which we come in contact every year, but we pass through the main body three times in a century — last in 1866 — capturing millions on each occasion. One of these has been graphically described by Humboldt :

"From half after two in the morning the most extraordinary luminary meteors were seen in the direction of the east. M. Bonpland, who had risen to enjoy the freshness of the air, perceived them first. Thousands of bodies and falling stars succeeded each other during the space of four hours. Their direction was very regular from north to south. They filled a space in the sky extending from due east 30° to north and south. In an amplitude of 60° the meteors were seen to rise above the horizon at east-north-east, and at east, to describe arcs more or less extended, and to fall towards the south, after having followed the direction of the meridian. Some of them attained a height of 40°, and all exceeded 25° or 30°. No trace of clouds was to be seen. M. Bonpland states that, from the

first appearance of the phenomenon, there was
not in the firmament a space equal in extent
to three diameters of the moon which was not
filled every instant with bolides and falling
stars. The first were fewer in number, but
as they were of different sizes it was impossible to fix the limit between these two classes
of phenomena. All these meteors left luminous traces from five to ten degrees in length,
as often happens in the equinoctial regions.
The phosphorescence of these traces, or luminous bands, lasted seven or eight seconds.
Many of the falling stars had a very distinct
nucleus, as large as the disc of Jupiter, from
which darted sparks of vivid light. The
bodies seemed to burst as by explosion ; but
the largest, those from 1° to 1° 15′ in diameter,
disappeared without scintillation, leaving behind them phosphorescent bands (trabes).
exceeding in breadth fifteen or twenty minutes. The light of these meteors was white.
and not reddish. which must doubtless be
attributed to the absence of vapour and the
extreme transparency of the air."[1]

[1] Humboldt, *Travels.*

The past history of the Leonids, which Le Verrier has traced out with great probability, if not proved, is very interesting. They did not, he considers, approach the Sun until 126 A.D., when, in their career through the heavens, they chanced to come near to Uranus. But for the influence of that planet they would have passed round the Sun, and then departed again for ever. By his attraction, however, their course was altered, and they will now continue to revolve round the Sun.

There is a remarkable connection between star showers and comets, which, however, is not yet thoroughly understood. Several star showers follow paths which are also those of comets. and the conclusion appears almost irresistible that these comets are made up of Shooting Stars.

We are told, indeed. that 150,000,000 of meteors. including only those visible with a moderate telescope. fall on the earth annually. At any rate, there can be no doubt that every year millions of them are captured by the earth, thus constituting an appreciable, and in the course of ages a constantly increasing, part of the solid substance of the globe.

THE STARS

We have been dealing in the earlier part of this chapter with figures and distances so enormous that it is quite impossible for us to realise them; and yet we have still others to consider compared with which even the solar system is insignificant.

In the first place, the number of the Stars is enormous. When we look at the sky at night they seem, indeed, almost innumerable; so that, like the sands of the sea, the Stars of heaven have ever been used as effective symbols of number. The total number visible to the naked eye is, however, in reality only about 3000, while that shown by the telescope is about 100.000.000. Photography, however, has revealed to us the existence of others which no telescope can show. We cannot by looking long at the heavens see more than at first; in fact, the first glance is the keenest. In photography, on the contrary,

no light which falls on the plate, however
faint. is lost; it is taken in and stored up.
In an hour the effect is 3600 times as great
as in a second.   By exposing the photographic
plate, therefore. for some hours, and even on
successive nights. the effect of the light is as
it were accumulated. and stars are rendered
visible. the light of which is too feeble to be
shown by any telescope.

The distances and magnitudes of the
Stars are as astonishing as their numbers,
Sirius. for instance. being about twenty times
as heavy as the Sun itself, 50 times as
bright. and no less than 1.000.000 times as
far away: while, though like other stars it
seems to us stationary. it is in reality sweeping through the heavens at the rate of 1000
miles a minute: Maia. Electra, and Alcyone,
three of the Pleiades. are considered to be
respectively 400. 480. and 1000 times as brilliant as the Sun. Canopus 2500 times. and
Arcturus. incredible as it may seem. even
8000 times. so that. in fact. the Sun is by
no means one of the largest Stars.   Even
the minute Stars not separately visible to the

naked eye, and the millions which make up
the Milky Way, are considered to be on an
average fully equal to the Sun in lustre.

Arcturus is, so far as we know at present,
the swiftest, brightest, and largest of all.   Its
speed is over 300 miles a second, it is said to
be 8000 times as bright as the Sun, and 80
times as large, while its distance is so great
that its light takes 200 years in reaching us.

The distances of the heavenly bodies are
ascertained by what is known as " parallax."
Suppose the ellipse (Fig. 54), marked Jan.,
Apr., July, Oct., represents the course of the
Earth round the Sun, and that A B are two
stars.   If in January we look at the star A,
we see it projected against the front of the
sky marked 1.   Three months later it would
appear to be at 2, and thus as we move round
our orbit the star itself appears to move in
the ellipse 1, 2, 3, 4.   The more distant star
B also appears to move in a similar, but
smaller, ellipse ; the difference arising from
the greater distance.   The size of the ellipse
is inversely proportional to the distance, and
hence as we know the magnitude of the

earth's orbit we can calculate the distance of
the star. The difficulty is that the apparent
ellipses are so minute that it is in very few
cases possible to measure them.

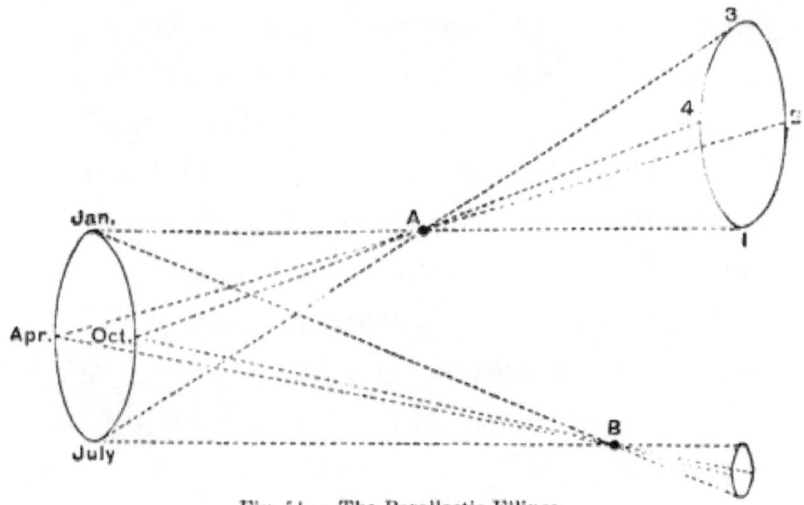

Fig. 54. — The Parallactic Ellipse.

The distances of the Fixed Stars thus tested
are found to be enormous, and indeed generally incalculable; so great that in most cases,
whether we look at them from one end
of our orbit or the other — though the difference of our position, corresponding to the
points marked January and July in Fig. 54,
is 185,000,000 miles — no apparent change of
position can be observed. In some, however,
the parallax, though very minute, is yet ap-

proximately measurable. The first star to
which this test was applied with success was
that known as 61 Cygni, which is thus shown
to be no less than 40 billions of miles away
from us — many thousand times as far as we
are from the Sun. The nearest of the Stars,
so far as we yet know, is a Centauri, the distance of which is about 25 billions of miles.

The Pleiades are considered to be at a distance of nearly 1500 billions of miles.

As regards the chemical composition of the
Stars, it is, moreover, obvious that the powerful engine of investigation afforded us by the
spectroscope is by no means confined to the
substances which form part of our system.
The incandescent body can thus be examined,
no matter how great its distance, so long only
as the light is strong enough. That this
method was theoretically applicable to the
light of the Stars is indeed obvious, but the
practical difficulties are very great. Sirius,
the brightest of all, is, in round numbers, a
hundred millions of millions of miles from us;
and, though as bright as fifty of our suns, his
light when it reaches us, after a journey of

sixteen years, is at most one two-thousand-millionth part as bright. Nevertheless, as long ago as 1815 Fraunhofer recognised the fixed lines in the light of four of the Stars ; in 1863 Miller and Huggins in our own country, and Rutherford in America, succeeded in determining the dark lines in the spectrum of some of the brighter Stars, thus showing that these beautiful and mysterious lights contain many of the material substances with which we are familiar. In Aldebaran, for instance, we may infer the presence of hydrogen, sodium, magnesium, iron, calcium, tellurium, antimony, bismuth, and mercury. As might have been expected, the composition of the Stars is not uniform, and it would appear that they may be arranged in a few well-marked classes, indicating differences of temperature, or perhaps of age.

Thus we can make the Stars teach us their own composition with light, which started from its source years ago, in many cases long before we were born.

Spectrum analysis has also thrown an unexpected light on the movements of the Stars.

Ordinary observation, of course, is powerless to inform us whether they are moving towards or away from us. Spectrum analysis, however, enables us to solve the problem, and we know that some are approaching, some receding.

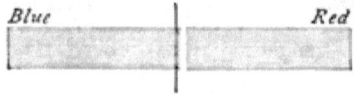

Fig. 55. — Displacement of the hydrogen line in the spectrum of Rigel.

If a star, say for instance Sirius, were motionless, or rather if it retained a constant distance from the earth, Fraunhofer's lines would occupy exactly the same position in the spectrum as they do in that of the Sun. On the contrary, if Sirius were approaching, the lines would be slightly shifted towards the blue, or if it were receding towards the red. Fig. 55 shows the displacement of the hydrogen line in the spectrum of Rigel, due to the fact that it is receding from us at the rate of 39 miles a second. The Sun affords us an excellent test of this theory. As it revolves on its axis one edge is always approaching and the other receding from us at a known rate, and observation shows that the lines given by the light of the two edges differ accordingly. So again as regards the Stars, we obtain a similar test derived from the Earth's movement. As we revolve in our orbit we approach or recede any given star, and our rate of motion being known we thus obtain a second test. The results thus examined have stood their ground satisfactorily, and in Huggins' opinion may be relied on within about an English mile a second. The effect of this movement is, moreover, independent of the distance. A lateral motion, say of 20 miles a second, which in a nearer object would appear to be a stupendous velocity, becomes in the Stars quite imperceptible. A motion of the same rapidity, on the other hand. towards or away from us, displaces the dark lines equally, whatever the distance of the object may be. We may then affirm that Sirius, for instance, is receding from us at the rate of about 20 miles a second. Betelgeux, Rigel, Castor, Regulus, and others are also moving away; while some — Vega, Arcturus, and Pollux. for example — are approaching us. By the same process it is shown that some groups of stars are only apparently in relation to one another. Thus in Charles' Wain some of the stars are approaching, others receding.

I have already mentioned that Sirius, though it seems, like other stars, so stationary that we speak of them as "fixed," is really sweeping along at the rate of 1000 miles a minute. Even this enormous velocity is exceeded in other cases. One, which is numbered as 1830 in Groombridge's *Catalogue of the Stars*, and is therefore known as "Groombridge's 1830," moves no less than 12,000 miles a minute, and Arcturus 22,000 miles a minute, or 32,000,000 of miles a day; and yet the distances of the Stars are so great that 1000 years would make hardly any difference in the appearance of the heavens.

Changes, however, there certainly would be. Even in the short time during which we have any observations, some are already on record. One of the most interesting is the fading of the 7th Pleiad, due, according to Ovid, to grief at the taking of Troy. Again, the "fiery Dogstar," as it used to be, is now, and has been for centuries, a clear white.

The star known as Nova Cygni—the "new star in the Constellation of the Swan"—was first observed on the 24th November 1876 by Dr. Schmidt of Athens, who had examined that part of the heavens only four days before, and is sure that no such star was visible then. At its brightest it was a brilliant star of the third magnitude, but this only lasted for a few days; in a week it had ceased to be a conspicuous object, and in a fortnight became invisible without a telescope. Its sudden splendour was probably due to a collision between two bodies, and was probably little, if at all, less than that of the Sun itself. It is still a mystery how so great a conflagration can have diminished so rapidly.

But though we speak of some stars as specially variable, they are no doubt all undergoing slow change. There was a time when they were not, and one will come when they will cease to shine. Each, indeed, has a life-history of its own. Some, doubtless, represent now what others once were, and what many will some day become.

For, in addition to the luminous heavenly bodies, we cannot doubt that there are countless others invisible to us. some from their greater distance or smaller size, but others. doubtless, from their feebler light : indeed. we know that there are many dark bodies which now emit no light. or comparatively little. Thus in the case of Procyon the existence of an invisible body is proved by the movement of the visible star. Again. I may refer to the curious phenomena presented by Algol. a bright star in the head of Medusa. The star shines without change for two days and thirteen hours; then in three hours and a half dwindles from a star of the second to one of the fourth magnitude : and then. in another three and a half hours. reassumes its original brilliancy. These changes led astronomers to infer the presence of an opaque body. which intercepts at regular intervals a part of the light emitted by Algol; and Vogel has now shown by the aid of the spectroscope that Algol does in fact revolve round a dark, and

therefore invisible, companion. The spectroscope, in fact, makes known to us the
presence of many stars which no telescope
could reveal.

Thus the floor of heaven is not only
" thick inlaid with patines of bright gold,"
but studded also with extinct stars, once probably as brilliant as our own Sun, but now
dead and cold, as Helmholtz tells us that our
Sun itself will be some seventeen millions of
years hence.

Such dark bodies cannot of course be seen,
and their existence, though we cannot doubt
it, is a matter of calculation. In one case,
however, the conclusion has received a most
interesting confirmation. The movements of
Sirius led mathematicians to conclude that it
had also a mighty and massive neighbour, the
relative position of which they calculated,
though no such body had ever been seen. In
February 1862, however, the Messrs. Alvan
Clark of Cambridgeport were completing
their 18-inch glass for the Chicago Observatory. "'Why, father,'" exclaimed the younger
Clark, "'the star has a companion.' The

father looked, and there was a faint star
due east from the bright one, and distant
about ten seconds. This was exactly the pre-
dicted direction for that time, though the dis-
coverers knew nothing of it. As the news
went round the world many observers turned
their attention to Sirius; and it was then
found that, though it had never before been
noticed, the companion was really shown under
favourable circumstances by any powerful
telescope. It is, in fact, one-half of the size of
Sirius, though only $\frac{1}{10000}$th of the brightness." [1]

Stars are, we know, of different magnitudes and different degrees of glory. They
are also of different colours. Most, indeed, are
white, but some reddish, some ruddy, some
intensely red; others, but fewer, green, blue,
or violet. It is possible that the comparative rarity of these colours is due to the fact
that our atmosphere especially absorbs green
and blue, and it is remarkable that almost all
of the green, blue, or violet stars are one of
the pairs of a Double Star, and in every case

---

[1] Clarke, *System of the Stars.*

the smaller one of the two, the larger being red, orange, or yellow. One of the most exquisite of these is β Cygni, a Double Star, the larger one being golden yellow, the smaller light blue. With a telescope the effect is very beautiful, but it must be magnificent if one could only see it from a lesser distance.

Double Stars occur in considerable numbers. In some cases indeed the relation may only be apparent, one being really far in front of the other. In very many cases, however, the association is real, and they revolve round one another. In some cases the period may extend to thousands of years; for the distance which separates them is enormous, and, even when with a powerful telescope it is indicated only by a narrow dark line, amounts to hundreds of millions of miles. The Pole Star itself is double. Andromeda is triple, with perhaps a fourth dark and therefore invisible companion. These dark bodies have a special interest, since it is impossible not to ask ourselves whether some at any rate of them may not be inhabited. In ϵ Lyræ there are two, each again being itself double.

ξ Cancri, and probably also θ Orionis, consist of six stars, and from such a group we pass on to Star Clusters in which the number is very considerable. The cluster in Hercules consists of from 1000 to 4000. A stellar swarm in the Southern Cross contains several hundred stars of various colours, red. green. greenish blue, and blue closely thronged together, so that they have been compared to a " superb piece of fancy jewellery." [1]

The cluster in the Sword Handle of Perseus contains innumerable stars, many doubtless as brilliant as our Sun. We ourselves probably form a part of such a cluster. The Milky Way itself, as we know. entirely surrounds us; it is evident, therefore, that the Sun, and of course we ourselves, actually lie in it. It is. therefore, a Star Cluster. one of countless numbers. and containing our Sun as a single unit.

It has as yet been found impossible to determine even approximately the distance of these Star Clusters.

[1] Kosmos.

NEBULÆ

From Stars we pass insensibly to Nebulæ, which are so far away that their distance is at present quite immeasurable. All that we can do is to fix a minimum, and this is so great that it is useless to express it in miles. Astronomers, therefore, take the velocity of light as a unit. It travels at the rate of 180,000 miles a second, and even at this enormous velocity it must have taken hundreds of years to reach us, so that we see them not as they now are but as they were hundreds of years ago.

It is no wonder, therefore, that in many of these clusters it is impossible to distinguish the separate stars of which they are composed. As, however, our telescopes are improved, more and more clusters are being resolved. Photography also comes to our aid, and, as already mentioned, by long exposure stars can be made visible which are quite imperceptible to the eye, even with aid of the most powerful telescope.

Spectrum analysis also seems to show that

such a nebula as that in Andromeda, which with our most powerful instruments appears only as a mere cloud, is really a vast cluster of stellar points.

This, however, by no means applies to all the nebulæ. The spectrum of a star is a bright band of colour crossed by dark lines; that of a gaseous nebula consists of bright lines. This test has been made use of, and indicates that some of the nebulæ are really immense masses of incandescent and very attenuated gas; very possibly, however, in a condition of which we have no experience, and arranged in discs, bands, rings, chains, wisps, knots, rays, curves, ovals, spirals, loops, wreaths, fans, brushes, sprays, lace, waves, and clouds. Huggins has shown that many of them are really stupendous masses of glowing gas, especially of hydrogen, and perhaps of nitrogen, while the spectrum also shows other lines which perhaps may indicate some of the elements which, so far as our Earth is concerned, appear to be missing between hydrogen and lithium. Many of the nebulæ are exquisitely beautiful, and their colour very varied.

In some cases, moreover, nebulæ seem to be gradually condensing into groups of stars, and in many cases it is difficult to say whether we should consider a given group as a cluster of stars surrounded by nebulous matter or a gaseous nebula condensed here and there into stars.

"Besides the single Sun," says Proctor, "the universe contains groups and systems and streams of primary suns; there are galaxies of minor orbs; there are clustering stellar aggregations showing every variety of richness, of figure, and of distribution; there are all the various forms of star cloudlets, resolvable and irresolvable, circular, elliptical, and spiral; and lastly, there are irregular masses of luminous gas clinging in fantastic convolutions around stars and star systems. Nor is it unsafe to assert that other forms and varieties of structure will yet be discovered, or that hundreds more exist which we may never hope to recognise."

Nor is it only as regards the magnitude and distances of the heavenly bodies that we are lost in amazement and admiration. The lapse of time is a grander element in Astronomy even than in Geology, and dates back long before Geology begins. We must figure to ourselves a time when the solid matter which now composes our Earth was part of a continuous and intensely heated gaseous body, which extended from the centre of the Sun to beyond the orbit of Neptune, and had, therefore, a diameter of more than 6,000,000,000 miles.

As this slowly contracted, Neptune was detached, first perhaps as a ring, and then as a spherical body. Ages after this Uranus broke away.

Then after another incalculable period Saturn followed suit, and here the tendencies to coherence and disruption were so evenly balanced that to this day a portion circulates as rings round the main body instead of being broken up into satellites. Again after successive intervals Jupiter, Mars, the Asteroids, the Earth, Venus, and Mercury all passed through the same marvellous phases. The time which these changes would have required must have been incalculable, and they all of course preceded, and preceded again by another incalculable period, the very commencement of that geological history which itself indicates a lapse of time greater than human imagination can realise.

Thus, then, however far we penetrate in time or in space, we find ourselves surrounded by mystery. Just as in time we can form no idea of a commencement, no anticipation of an end, so space also extends around us, boundless in all directions. Our little Earth revolves round the mighty Sun; the Sun itself and the whole solar system are moving with inconceivable velocity towards a point in the constellation of Hercules; together with all the nearer stars it forms a cluster in the heavens, which appears to our eyes as the Milky Way; while outside our star cluster again are innumerable others, which far transcend, alike in magnitude, in grandeur, and in distance, the feeble powers of our finite imagination.